The Future of Airbnb and the 'Sharing Economy'

THE FUTURE OF TOURISM

Series Editors: Ian Yeoman, *Victoria University of Wellington, New Zealand* and Una McMahon-Beattie, *Ulster University, Northern Ireland, UK*

Some would say that the only certainties are birth and death; everything else that happens in between is uncertain. Uncertainty stems from risk, a lack of understanding or a lack of familiarity. Whether it is political instability, autonomous transport, hypersonic travel or peak oil, the future of tourism is full of uncertainty, but it can be explained or imagined through trend analysis, economic forecasting or scenario planning.

This new book series, The Future of Tourism, sets out to address the challenges and unexplained futures of tourism, events and hospitality. By addressing the big questions of change, examining new theories and frameworks or critical issues pertaining to research or industry, the series will stretch your understanding and generate dialogue about the future. By adopting a multidisciplinary perspective, be it through science fiction or computer-generated equilibrium modelling of tourism economies, the series will explain and structure the future – to help researchers, managers and students understand how futures could occur. The series welcomes proposals on emerging trends and critical issues across the tourism industry and research. All proposals must emphasise the future and be embedded in research.

All books in this series are externally peer-reviewed.

Full details of all the books in this series and of all our other publications can be found on http://www.channelviewpublications.com, or by writing to Channel View Publications, St Nicholas House, 31–34 High Street, Bristol BS1 2AW, UK.

THE FUTURE OF TOURISM: 1

The Future of Airbnb and the 'Sharing Economy'

The Collaborative Consumption of our Cities

Jeroen A. Oskam

CHANNEL VIEW PUBLICATIONS
Bristol • Blue Ridge Summit

DOI https://doi.org/10.21832/OSKAM6737
Library of Congress Cataloging in Publication Data
A catalog record for this book is available from the Library of Congress.
Names: Oskam, Jeroen A. – author.
Title: The Future of Airbnb and the 'Sharing Economy': The Collaborative
   Consumption of our Cities/Jeroen A. Oskam.
Description: Bristol, UK; Blue Ridge Summit, PA: Channel View Publications, 2019. |
   Series: The Future of Tourism: 1 | Includes bibliographical
   references and index.
Identifiers: LCCN 2018051443 | ISBN 9781845416737 (hbk : alk. paper) |
   ISBN 9781845416720 (pbk : alk. paper) | ISBN 9781845416768 (kindle)
Subjects: LCSH: Peer-to-peer travel—Economic aspects. | Airbnb (Firm)
Classification: LCC G156.5.P44 O75 2019 | DDC 910.68—dc23 LC record available
   at https://lccn.loc.gov/2018051443

British Library Cataloguing in Publication Data
A catalogue entry for this book is available from the British Library.

ISBN-13: 978-1-84541-673-7 (hbk)
ISBN-13: 978-1-84541-672-0 (pbk)

Channel View Publications
UK: St Nicholas House, 31–34 High Street, Bristol, BS1 2AW, UK.
USA: NBN, Blue Ridge Summit, PA, USA.

Website: www.channelviewpublications.com
Twitter: Channel_View
Facebook: https://www.facebook.com/channelviewpublications
Blog: www.channelviewpublications.wordpress.com

Copyright © 2019 Jeroen A. Oskam.

All rights reserved. No part of this work may be reproduced in any form or by any means without permission in writing from the publisher.

The policy of Multilingual Matters/Channel View Publications is to use papers that are natural, renewable and recyclable products, made from wood grown in sustainable forests. In the manufacturing process of our books, and to further support our policy, preference is given to printers that have FSC and PEFC Chain of Custody certification. The FSC and/or PEFC logos will appear on those books where full certification has been granted to the printer concerned.

Typeset by Nova Techset Private Limited, Bengaluru and Chennai, India.
Printed and bound by CPI Group (UK) Ltd, Croydon, CR0 4YY
Printed and bound in the US by Thomson-Shore, Inc.

Contents

	Introduction	1
	Background	1
	Two Isolated Debates	2
	The Platforms, their Clients and their Externalities	3
	The Future of Urban Vacation Rentals	5
1	The 'Sharing' Debate	7
	The Sharing Utopia	7
	For-profit Platforms	12
	'Sharing' as a Push for Deregulation	16
	Extraction	18
	Summary: Neither Underutilised Assets nor Human Connection	19
2	Platforms or Two-sided Markets	20
	Two-sided Markets	20
	Winner-takes-all Competition	21
	Online Travel Agents	23
	Urban Vacation Rental Platforms	24
	Booking.com: The Threat of Envelopment by Other Platforms	26
	The Napster Scenario	27
	Blockchain	28
	Summary: Oligopolistic Market Power	30
3	Living Like a Local! But How Do Locals Live?	31
	Authenticity	31
	A Community of True Believers	33
	Airbnb Promotion Campaigns	36
	Authenticity is Bliss	40
	Summary: Cult Admission	41
4	Millennial or Mainstream: The Airbnb Guest	42
	Motivations to Participate in Sharing	43
	Guest Satisfaction	47
	Airbnb for Business	48
	Further Research into Airbnb Guest Demographics	49
	Summary: Price, Space and Amenities	51

5	Adventurous Start-up or Real-estate Tycoon: The Airbnb Host	52
	Hosts' Profile	53
	Multilisters	54
	Professionalisation	58
	Concierge Companies	59
	Working Relations in the 'Sharing Economy'	60
	Entrance Barriers	62
	The Work of a Host: Setup, 'Overheads' and Booking-related Tasks	63
	The Empowerment of Poor Residents	64
	Summary: Free Market Equality	65
6	The Established Order: Impact on the Hotel Market	67
	Market Share	68
	Hotel Companies' Competitive Strategies	70
	Airbnb Entering the Hotel Business	73
	Airbnb as a Distribution Platform	74
	Summary: Part of the Industry	78
7	Suburb Safari or Hotspot Hype? Airbnb in the Neighbourhoods	79
	Spatial Concentration	80
	Gentrification and Overtourism	85
	Urban Vacation Rentals and the Housing Market	88
	Summary: The Commodification of Neighbourhoods	92
8	The Regulation of Urban Vacation Rentals: Empowered Residents, Emasculated Authorities?	93
	Fragmentation	94
	Reasons to Regulate	95
	Existing Regulation: Acquiescence or Prohibition	96
	Specific 'Sharing' Regulation	97
	Regulation Enforcement	99
	A Fight for Transparency	102
	Summary: Compliance and Transparency Stifle Growth	104
9	What if This Does Not Stop? Amsterdam 2025	105
	'A Great and Unique Approach'	106
	Triple-digit Growth	108
	The 'Venice of the North'	110
	Scenario 1: A Platform Complements the Hotel Offer in a 'Touristified' City	113
10	What if the City Fights Back? Barcelona 2025	114
	'Neighbours: A Species Threatened with Extinction'	115
	Anti-tourism Protests	116
	Political Turmoil	117

	Barcelona, 'the Pebble in Airbnb's Shoe'	118
	Can a Repressive Policy Be Successful?	119
	Fighting the *Easyjetset*: A New Monte Carlo?	120
	Scenario 2: Outlawing the Platforms and Deterring Low-cost Tourists	121
11	What if the City Regains Control? San Francisco 2025	122
	The San Francisco Housing Crisis	124
	San Francisco Tourism Statistics	128
	Genuine Sharing	128
	A Grassroots Counter-movement?	130
	Scenario 3: Commercial, Hotel-like Rentals Versus a Revival of Genuine 'Sharing'	132
12	What if the City Outsmarts the Market? Singapore 2025	133
	Tourism in Singapore	134
	Airbnb Regulations	136
	Smartness	137
	Controlled 'Sharing'	138
	Scenario 4: Platform Hospitality Under Surveillance	139
13	Conclusion: A Neoliberal Nightmare	141
	Debunking the Platforms' Narrative	142
	A Grim Future: Why is There No Positive Scenario?	143
	The Commodification of Everything	144
	Future Outlook: Our Cities	147
	Future Outlook: Vacation Rental Platforms	148
	The Future of Commodified Tourism	149
References		151
Index		172

Introduction

Background

The interest in accommodation 'sharing' and specifically Airbnb was born a few years ago, and first took shape with a scenario workgroup composed of practitioners and academics at the Amsterdam campus of Hotelschool The Hague in 2015. At that time, home 'sharing' was still widely considered as a marginal curiosity in Amsterdam, and probably in many other places in Europe as well. This was initially also the predominant perspective in the scenario workgroup. Most participants wondered ingenuously about the appeal of sleeping in a stranger's house in some neighbourhood, hotel companies were looking for ways to incorporate a similar kind of authenticity, while another concern was whether in the future it could become a competitive threat to hotel businesses.

A first reality check came when one of the participants – a tourism researcher from a renowned national organisation – asked if anyone knew how many visitors we had in Amsterdam because of Airbnb. Strange as it may sound, the phenomenon still seemed so limited at that time that nobody had given much thought to the volumes. This would be completely unthinkable nowadays, but the context was also different: in a continent just recovered from an economic and financial crisis, tourism growth was still welcomed as it generated jobs and spurred consumer confidence. At the same time, Amsterdam had proclaimed itself a 'sharing city', as an anti-consumerist and emancipatory answer to exactly the inefficacies and inequalities that had surfaced in the previous years.

Only when we started digging deeper, looking for numbers, did the stench of actual 'sharing' practices and their externalities become stronger. We found that Airbnb did not facilitate any independent research. The only numbers available were data scraped from the website by activist researchers. Murray Cox and Tom Slee revealed the inequalities in the practices of Airbnb and the misrepresentations of their marketing narrative. Their studies opened our eyes to the commercial speculation – to the detriment of neighbourhoods and residents – that had emerged in what we stopped calling 'sharing'. In the 'Sharing City' of 2015, where progressive politicians in particular still identified with that narrative, this perspective was totally heretic. Our scenarios envisioned a commercialisation of the Airbnb offer in Amsterdam, which at that time sounded outrageous.

A partnership with the research department of Colliers International enabled us to purchase AirDNA data scrapes. The advantage of these data is that they are exhaustive. Whereas most scrapes are snapshots of the Airbnb offer at a specific moment, AirDNA's daily scrapes allowed for a more complete and detailed analysis, since the offer of urban vacation rentals is extremely volatile: listings are not always visible, but taken off and put back on the market. Also, AirDNA worked with a machine learning algorithm that gives insight into actual rental activity which since 2014 had been concealed in Airbnb's API: the data included a reservation status proxy which was extrapolated from the length of the new unavailable period, the time from the unavailable period from the last change, and the historical behaviour of the property. This implies a limitation, but still the AirDNA data constitute the most accurate approximation available of Airbnb's performance.

The Airbnb city reports that we subsequently started publishing with Colliers confirmed the profound changes in the market precisely in the course of this first scenario exercise. From 18,787 booked nights that were detected in January 2015, demand exploded to 107,418 nights one year later, a growth of 474%. This seemed extreme, but our outcomes could be triangulated with other data. With the 736,000 total booked nights, divided in a simple formula by the average length of stay (of 3.69 nights for Amsterdam during the entire year), and multiplied by an estimated party size (of between 2.52 and 3.00) per booking, we were able to calculate an approximate visitor number to Amsterdam of 555,000 for 2015. This outcome was corroborated by a survey-based study of NBTC Holland Marketing and by the tourist tax declaration of Airbnb to the city, which spoke of 575,000 visitors (Amsterdam Marketing, 2016; Bakker *et al.*, 2016a; Vranken & Van der Most, 2016). In 2016, we found 1.1 million visitors to Amsterdam, plus 88,000 for The Hague and 90,000 for Rotterdam, again in line with NBTC (Bakker *et al.*, 2017a, 2017b; NBTC Holland Marketing, 2017). Although Airbnb did not publish any numbers for that year and rejected our study, the numbers were also corroborated by a press release by the platform itself that spoke of 1.4 million visitors to the entire country (Sondermeijer, 2017). Our calculations for 2017 – 1.4 million visitors to Amsterdam – could not yet be contrasted with any other numbers (Bakker *et al.*, 2018a).

Two Isolated Debates

These staggering growth numbers surprised us, just as they surprised our society and our politicians. Why had nobody warned them? Unfortunately, the academic contribution to the social debate has remained mostly sterile. At this early stage in the development of urban vacation rentals, only limited literature was available: the most noteworthy were preprints of Guttentag (2015) and Zervas *et al.* (2017), two articles by Stors and Kagermeier (2015; Kagermeier *et al.*, 2015), and some studies of specific aspects, such as those of Ikkala and Lampinen (2014). In the case of a rapidly evolving phenomenon, the lead time of academic articles – for peer-review and production – makes them almost obsolete by the time they are published. In a world driven by research metrics, it can become fatal to share socially relevant data before publication.

It is even more disappointing that, now that we have recently seen a true avalanche of studies on urban vacation rentals, so many of them follow business study templates that contribute little to the most pressing concerns in society. How can it make sense for an article to speak of 'sharing', without any critical questioning of the concept, and then proceed to formulate elaborate marketing or pricing advice to supposedly dilettante Airbnb hosts in less privileged neighbourhoods? When society urgently needed a conceptual discussion on the nature and impact of these phenomena, research paradigms drove academia towards a sterile and repetitive exercise that could only recognise and reinforce the money-making schemes that had become prevalent in 'sharing'.

In the meantime, the public debate shifted from an initial enthusiasm for the social and environmental benefits of 'sharing' to its negative externalities, for the labour market, mainly in the case of Uber, and for housing and tourist pressure on cities, in the case of Airbnb. Very generally, and without denying the existence of more lucid and better informed positions in the debate, 'sharing' was initially seen as a progressive principle and critical attitudes were deemed politically and socially conservative, contrary to innovation, before the shift in the debate made these positions more diffuse and more confused. This misunderstanding has been instrumental to the acceptance and growth of Airbnb and other platforms.

Whereas the academic debate remained largely sterile, the public debate became aggressive, with voracious verbal attacks and insults aimed at those who criticised the 'sharing' platforms. The aggression can only be explained by the important financial stakes some had in Airbnb, by the responsibility of embarrassed politicians who had let it all happen, by the fanaticism of 'sharing' zealots and by the nervousness of those who became confused in the shifting debate. The Airbnb PR apparatus accused all its critics of being part of a big hotel-lobby conspiracy.

The Platforms, their Clients and their Externalities

This book wants to contribute to the debate on the nature of urban vacation rentals such as Airbnb, HomeAway or Wimdu, on their impact and on their future evolution. Rather than by taking a programmatic stand on the good and evil of this phenomenon, it does so by collecting and synthesising the – quite heterogeneous – information that is now available in academic literature and in events reflected in the news, supported by findings from our own data analyses wherever possible and relevant. One of our objectives is bridging the gap between the academic and the public debate.

The first chapter discusses the concept of 'sharing' itself. The word 'sharing' has become like 'freedom': it has so often been used to defend its exact opposite that now there is hardly any common understanding of what it means. To avoid ideological denominations, I will try to stick to the most factual description of what is happening: people renting out city apartments to tourists, hence 'urban vacation rentals'. Airbnb has, like Kleenex or Pop-Tarts, become a genericised trademark for services sold under different brands. It is also, as market leader, the source of most examples; however, to avoid ambiguity the name 'Airbnb' will only be used to refer to that specific platform.

In subsequent chapters, different aspects of vacation rentals will be examined. Services like Airbnb generally describe themselves as communities, which reflects their utopian 'sharing' inspiration but not their actual organisation. To understand the growth dynamics of urban vacation rentals, it is key to discuss the matchmaking role they have as intermediary platforms. This explains the emergence of a 'winner-takes-all' competition, in which Airbnb is ahead thanks to a very successful marketing strategy. Why is the 'living like a local' message so successful, and how has the company convinced people that, even when they travel and end up gazing at the Eiffel Tower, they are still not tourists?

Analysing the clients of Airbnb – both travellers who stay in vacation rentals ('guests'), and the supply side of residents or owners (so-called 'hosts') – is difficult without access to data. Platforms have three reasons to be reluctant to give this insight. In the first place, the identity of their clients and their behaviour is of course commercially valuable information they will not share with competitors. But also the questionable legal status of rental activity often makes this information sensitive: Are their hosts paying taxes? Are they illegally subletting? Are they allowed to provide accommodation at all? Finally, the platforms carefully protect an image of their client 'communities' which may not correspond to their actual practice. While policy makers and commercial parties often still rely on this marketing narrative, it is important to verify if this really is the grassroots movement of adventurous millennials and ordinary residents struggling to make ends meet, or if it has become a more mainstream commercial activity.

For the external consequences of the evolution of urban vacation rentals, there are two urgent questions. The first concerns how the phenomenon has changed the competitive forces in the hotel and tourist industry. Do urban vacation rentals generate incremental demand or are they taking market share from traditional accommodation providers? What will be the next strategic moves of the rental platforms and of other companies?

With the exponential growth of urban vacation rentals, their external effects on cities, neighbourhoods and residents have started to dominate the public debate. Next to the triple transaction between 'hosts', 'guests' and platforms extracting a commission, third parties (fourth, properly speaking) pay a price through increased tourism traffic, housing market effects and other changes to their environment. What are these effects, and is Airbnb the disease or, as the company claims, the cure for residents' problems?

In the study of these topics, I have tried to keep up with the stream of literature that has been abundant in the last two years. The fact that an article or book is not cited does not imply any judgement on the relevance or quality of that work; this is an ongoing and rapidly evolving debate, and there will probably have been important involuntary omissions from my side.

The Future of Urban Vacation Rentals

Understanding what 'sharing' means, how its market works, the identity and motivations of its customers and the impact of all this on external parties allows us to think about how urban vacation rentals may further evolve. The future is a function of the present and if only we knew all the variables it would be easy to predict further developments. Unfortunately, we don't, or as Hines and Bishop (2013) expressed it, 'A funny thing happened on the way to the baseline future – something else'. Hence the general preference for outlining multiple plausible alternatives for future developments or scenarios.

This is not the place for an in-depth discussion of the wide array of scenario planning methods (Bishop *et al.*, 2007), but the most well-known and widest used technique is deducing different options from the main uncertainties on our way to the future. This Global Business Network (GBN) approach is especially appropriate for elaborating coherent narratives based on the diverse input of multiple stakeholders (Van der Heijden, 2005). We can also inductively arrive at different scenarios by studying as complete as possible a range of variables and then modifying their basic assumptions, as in the Framework Foresight Method (Hines & Bishop, 2013). Rather than moderating stakeholders' views and opinions, this method helps makes sense of multiple research and policy inputs (Oskam & Zandberg, 2016).

In this desk research exercise, my first approach has been to identify most of the currently known variables in order to arrive at an understanding of a 'baseline' future, or 'the fundamental future with no surprises' (Hines & Bishop, 2013: 37). To design four radically different alternatives, I have used a variation of this method, changing the baseline assumptions according to the Four Futures archetypes proposed by Dator (2009): 'continued growth', 'collapse', 'disciplined society' and 'transformational society'. Without pretending to achieve any predictive accuracy, these alternative courses of development help imagine the consequences of different social and political responses to the urban vacation rental phenomenon. The imagined scenarios are placed in four different contexts, not as specific outlooks for those cities but rather as the result of responses that might plausibly occur there.

After the development of four scenarios based on this distinction, namely, further growth of the urban vacation rental phenomenon versus its extinction, a 'disciplined' interpretation in the sense of a return to its original intentions versus a technological 'transformation' of society that allows it to manage the phenomenon, in fact two scenario axes became distinguishable as uncertainties leading to those different outcomes. The first two scenarios are more likely in a situation where urban vacation rentals keep evolving in their current secrecy, whereas the latter two seem more probable if, one way or another, the platforms are forced into greater transparency. The 'growth' and 'discipline' scenarios will occur if vacation rentals are allowed or legalised, whereas the 'collapse' and 'transformation' will most probably be accompanied by a legal crackdown on vacation rental platforms.

There is no preferred nor worst-case scenario; the envisioned scenarios should not be seen as policy recommendations either. They should be read as imagined outcomes of well-informed analyses of the evolution of urban vacation rentals up to 2018, meant to contribute to the conceptual debate on what the so-called 'sharing' phenomenon is, whom it benefits and how big it can grow. In particular, if we examine this trend in the wider context of social and ideological developments, where will it lead? Will it have the empowering and emancipatory effects that some have ascribed to it, or are we heading in a different direction?

Acknowledgements

Airbnb has systematically accused critical voices of representing a commercial hotel lobby. I work for a state-funded higher education institute, which makes my work fully sponsored by the Dutch taxpayer. I can confidently verify that neither I, nor the leftist City Council of Barcelona, nor the California Federation of Teachers, are part of the hospitality industry.

Airbnb, however, is.

1 The 'Sharing' Debate

> Sharing is 'commerce with the promise of human connection'.
> Joe Gebbia, CFO Airbnb, TedX, 2016

The Sharing Utopia

Despite the definitions of the platform economy being widely debated, urban vacation rentals are still popularly considered as an example of 'sharing', with Airbnb often regarded as the most emblematic manifestation of the 'sharing economy'. Even in the academic world it is common to include the topic of Airbnb in publications and meetings that discuss the 'sharing' phenomenon. In its self-studies and press releases, the company underscores the emancipatory effects vacation rentals allegedly have, as well as the environmental benefits compared to traditional accommodation. It seems evident that the 'utopian spin' has contributed to the image of the platform and, therefore, to its explosive growth.

The central idea of 'sharing' was defined by Rachel Botsman (2010) as 'the more efficient use of underutilized assets'. The ownership paradigm makes us buy goods we do not intensively use. The typical example is a power drill which, according to Botsman, will be used on average during 12 minutes of its entire lifetime. Something similar can be said of cars, which after bringing us to work stand idle for eight hours until we go home. The ideal solution would be to give others access to these goods when we are not using them: not only would this save resources, but it would also bring these goods within reach of those unable to purchase them. This emancipatory effect is ultimately incompatible with for-profit models, as greater efficiency is achieved through a reduction of consumption.

Even though human interaction does not flow inevitably from the economics of access, the usual practice of jointly using an asset, as well as the utopian connotations of the movement promoting it, make the community aspect part of 'sharing' by association. It should be noted, though, that these are two different functions of 'sharing'. Therefore, if Joe Gebbia (2016) acknowledges that this platform does one, but not the other – by refining 'sharing' as 'commerce with the promise of human connection' – we have to understand that 'sharing' is stripped of its objectives of greater efficiency and broader access.

However, there is a more fundamental problem that sets vacation rentals apart from 'sharing' initiatives. If a user shares his or her car, its use and value will not be altered since for the other user it will still be a means of transportation. In a similar way, at moments when the owner of a power drill has no need for drilling holes, someone else can use the tool but only for the exact same purpose of drilling holes. However, if I temporarily give up my urban housing, this does not mean that I do not need housing nor, most of the time, that the temporary user will use my residence as housing. In fact, since the exchange value of tourist accommodation exceeds that of housing, a resident will have a strong incentive to change the function of that good. So, instead of multiplying the access to a scarce good by reducing its idle time, the availability of the good itself is reduced.

When this type of internet-enabled networked exchange and transaction first emerged, it was of course unclear how the phenomenon would evolve. Some early publications on 'sharing', 'collaborative consumption' or 'peer to peer' should therefore be considered as programmatic pamphlets about where the movement should head, rather than as academic studies of past practices. 'Embedded' authors, who combine their scholarly work with some kind of 'sharing' entrepreneurship, are no exception among the advocates of the phenomenon (Dolnicar, 2018; Sundararajan, 2016). Eventually, the actual development of the initiatives could therefore be deemed 'authentic' examples of the initial intentions or, for instance, commercial detractions. In fact, in his analysis of the early sharing discourse, Martin (2016) considers these two alternatives as potential pathways for the evolution of 'sharing': one of them would be aligned with a transition to sustainability, while the second one would reinforce the current unsustainable economic paradigm. At this early stage, his study identifies different ways in which the phenomenon was framed: as an economic opportunity, a more sustainable form of consumption, a pathway to a decentralised, equitable and sustainable economy or, on the contrary, as creating unregulated marketplaces, reinforcing the neoliberal paradigm or as an incoherent field of innovation.

The distinction between a prescriptive or a descriptive approach is not completely clear in all definitions (for an elaborate overview, see Gyimóthy & Dredge, 2017). Do they refer to what 'sharing', etc., should become, or do they analyse its actual manifestations? It is therefore understandable that an array of different definitions has been applied to urban vacation rental platforms. We need to carefully analyse the implications of each of these denominations in order to understand how vacation rentals relate to other services and initiatives in the 'sharing' movement or phenomenon.

Sharing

The concept of 'sharing' refers to common, in-group use of a good without the need to reciprocate – or, in Belk's definition, 'the act and process of distributing what is ours to others for their use and/or the act and process of receiving or taking something from others for our use' (Belk, 2010: 717). To distinguish sharing from exchanging commodities and gift giving, Belk proposes prototypes as more elucidating than definitions of the problematically delimitable characteristics of each act: mothering or the pooling and allocation of household resources in a family are prototypes of 'sharing', while buying a loaf of bread in a store is a typical 'commodity exchange'. Giving gifts is in appearance disinterested but supposes a reciprocity as precisely the mechanism that establishes a bond between giver and receiver.

Thus, a shared good is jointly used either inside a family or in a wider community as a manifestation of the 'extended self'. Family members eating the household food or villagers meeting in a public square do not incur specific debts or other obligations, besides the general duties and taxes they have as community members. These acts or attitudes towards goods are referred to as 'sharing in'; Belk speaks of 'sharing out' if something is used jointly with relative strangers. Neighbours splitting the cost of a roof repair or a lawn mower could be an example of 'sharing out', which would mean that they divide resources among their discrete economic interests, without this pragmatic union affecting their boundary between 'self' and 'other': once the roof is repaired they will not necessarily have established a different level of intimacy (Belk, 2010, 2014).

Access

'Access' does not have similar altruistic connotations; on the contrary, for Rifkin (2000: 182) the shift from a regime of property to one of access marks the 'commodification of lived experience'. We are entering, according to this author, a stage of hypercapitalism in which – as his subtitle states – 'all of life is a paid-for experience'. Afterwards, authors such as Botsman (2010) and Gansky (2010) put the term in a more positive light, equating it to 'sharing', reducing it to a less materialistic alternative to ownership. Thus, 'access' has become commonly accepted as offering ethical and environmental advantages over owning assets.

As in the case of the shared lawn mower, this 'sharing out' creates greater efficiency by making goods available at lower prices, albeit not 100% of the time: there is a trade-off between the full control of ownership and the temporary access of the shared user. In these cases, the access results from a transaction between supplier and user, and not from the common interest of a 'sharing' community. Work on car sharing shows that a company such as Zipcar was unsuccessful in promoting its user community because its users felt little engagement to their peers and mainly sought cheap and convenient access to the service (Bardhi & Eckhardt, 2012; Eckhardt & Bardhi, 2015).

Collaborative consumption

The latter interpretation of 'access' or 'sharing out' is also referred to as 'collaborative consumption' (Botsman, 2010). This denomination implies an empowerment of consumers in achieving access to the desired assets. Consumer solidarity, which manifests itself as a 'collaborative lifestyle' (Botsman & Rogers, 2011), turns efficiency-inspired pragmatism into an idealist movement. The idea of strangers joining hands to consume together reintroduces the idea of 'community', paradoxically linked to 'sharing out'.

Peer-to-peer

Even though Bauwens *et al.* (2012) still seem to use 'sharing', 'collaborative consumption' and 'peer-to-peer' (P2P) interchangeably, the term implies a different role for the user community from just jointly accessing existing assets, as it clearly intends to describe transactions between users. In other words, users collaborate to produce and share use-value. This means that P2P processes occur in 'distributed networks', defined as 'networks in which autonomous agents can freely determine their behavior and linkages without the intermediary of obligatory hubs' (Bauwens, 2005).

This ideal form of collaboration is at odds with the intermediary role of platforms, especially if these appropriate value generated by the users. These 'extractive' platforms thus organise 'a more parasitical form of capitalism, in which capital no longer organizes production, but facilitates P2P exchanges' (Araya, 2015; Bauwens, 2014).

Networked platforms

The distinction between peers who become connected to communicate, barter and engage in other types of transactions, and the mediated connections that facilitate those collaborative actions, is fundamental to understanding the economic forces at work. The simultaneous emergence of different applications for P2P connections, as well as the ambiguity of the word 'sharing' in its common usage, has created the illusion of similarity. Strictly speaking, this similarity is restricted, however, to the fact that these applications are facilitated by internet connectivity. But if we share a family picture on Facebook, a book via BookCrossing or a ride on Uber, we are talking about different acceptations of 'sharing' that cannot be captured in a single definition.

It is therefore necessary to analyse how peers are connected and how their collaborative actions relate to – are facilitated by and/or benefit – the connecting agent. In the narrower context of vacation rentals, the different types of P2P interaction can be illustrated by comparing Couchsurfing and Airbnb. Couches are often underutilised where people sleep; also, a couch surfer inevitably enters into contact with the host. This makes the activities on the platform, which is intended for non-monetary exchanges, genuine 'sharing' according to Botsman's and Gebbia's definitions. The motivation for users – both hosts and guests – to engage in sharing are based on a 'moral economy' aligned with that of alternative tourism in general: anti-consumerist travel, establishing real contact with the host and the destination and even 'undermining the corporate hotel machine' (Molz, 2013: 225). Couchsurfing itself, at least in its original form, is a not-for-profit organisation inspired by the ideals of the early internet as a facilitator of free communication. The Airbnb rhetoric mimics all these motivations; nevertheless, the platform is geared towards hosts and guests performing monetary transactions and the company itself extracting a percentage. As we will see in Chapters 4 and 5, some Airbnb users certainly invoke anti-consumerism and human contact, but this ideological driver is in general overshadowed by financial motives.

Boswijk proposes to clarify the role intermediary platforms play with a typology that establishes a distinction along two dimensions, as shown in Figure 1.1: the commons versus the private/commercial on the horizontal axis, and open versus controlled and closed systems on the vertical axis. Boswijk thus identifies four types of value networks (adapted from Bauwens, 2014; Kostakis & Bauwens, 2014):

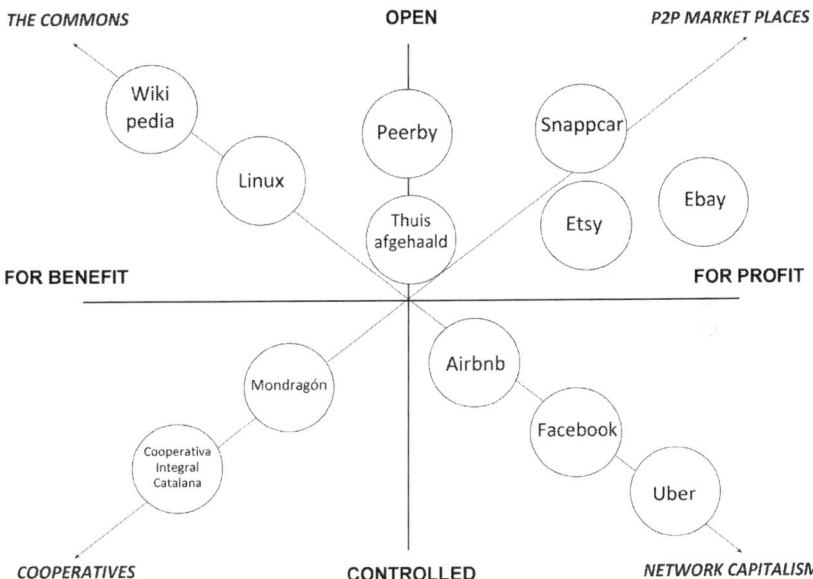

Figure 1.1 Boswijk's typology of value networks
Source: Oskam and Boswijk (2016).

(1) In the upper left quadrant, open and not-for-profit systems, such as Wikipedia or Linux.
(2) In the upper right quadrant, P2P social marketplaces based on open systems and with a fine-tuned distributed market function.

(3) In the lower left quadrant, collectives that are characterised through a closed protected system and for the common good.
(4) In the lower right quadrant, network capitalists, based on hyperconnected and distributed platforms with a commercial goal (Oskam & Boswijk, 2016).

For-profit Platforms

Can the data on Airbnb performance shed further light on whether urban vacation rentals can somehow be considered as 'sharing', or rather as a purely commercial activity? For governmental bodies, in order to determine whether an activity is professional or of a different nature, the criterion usually depends on fiscal regulations on the significance of the generated revenues as a share of a subject's income. This would be a strict criterion that might lead to a majority of Airbnb rentals qualifying as a professional activity, if we consider that the average annual revenue of a host in Amsterdam is €10,600 per unit (Bakker et al., 2018a). However, the limited availability of this type of data – for taxation purposes – is one of the concerns of city destinations and national tax agencies with the phenomenon. We will therefore concentrate on more conservative criteria concerning the characteristics of the offer.

The type of listing offered on Airbnb – shared room, private room or entire home or apartment – is an important indication of the commercial or 'shared' nature of the activity, as Slee argues (2015). This is as relevant for the non-materialist, efficiency-driven interpretation of 'sharing' as for the community aspect, which seems closer to Airbnb's 'commerce with the promise of human connection'. In the first place, if a host has room to spare in his or her own residence, that space could indeed be seen as an 'underutilised asset'. If a host has to abandon his or her home, or buys a property for the purpose of offering it on Airbnb, we can of course not speak of 'idling' capacity: in these cases, additional capacity is required. In the second place, a host receiving guests in a house where he or she also stays will of course offer a better opportunity for contact between locals and tourists.

The numbers in Table 1.1 show that 'shared rooms' are marginal in the Airbnb offer in these five cities. The majority of listings and of reservation nights are in entire homes, with the total revenue for 2015 close to or well over 80%. The predominance of entire homes is strongest in Paris and Amsterdam, where the vast majority of the offer is of this type.

However, there is a situation in which entire home rentals should be considered as the 'sharing' of an underutilised asset: this is the case if the residents are temporarily absent, for instance because of a holiday. Such a transaction would be similar to a traditional home swap. This makes it relevant to calculate the days units are offered on Airbnb. If a house is rented out or available for booking, the residents must have an alternative; therefore, the availability of units used for these 'home swaps' would normally be expected to be less than a month. Table 1.2 shows how many

Table 1.1 Type of listing, active listings 2017,[a] Amsterdam, Berlin, London, Madrid, Paris

	Amsterdam				Berlin				London				Madrid				Paris			
	Listings	Listings %	Booked nights %	Revenue %	Listings	Listings %	Booked nights %	Revenue %	Listings	Listings %	Booked nights %	Revenue %	Listings	Listings %	Booked nights %	Revenue %	Listings	Listings %	Booked nights %	Revenue %
Shared room	248	0.6	0.4	0.3	939	1.7	1.0	0.5	2,125	1.4	0.9	0.2	709	2.0	0.9	0.3	1,444	1.1	0.9	0.3
Private room	8,457	21.9	30.8	21.2	24,607	45.3	44.3	27.1	73,207	46.8	41.4	20.2	14,514	41.9	29.6	13.2	15,974	12.3	10.5	6.1
Entire home/apt.	29,890	77.3	68.7	78.4	28,691	52.8	54.6	72.4	79,770	51.0	57.5	79.4	19,331	55.8	69.3	86.3	112,501	86.4	88.5	93.5
Unknown/other	56	0.1	0.1	0.1	121	0.2	0.1	0.1	1,228	0.8	0.2	0.2	116	0.3	0.3	0.2	223	0.2	0.1	0.1
Total	38,651				54,358				156,330				34,670				130,142			

Notes: [a] Airbnb listings are considered 'active' if they are available (i.e. they are either booked, or can be booked), or if they have had calendar updates in the previous two months, or if they have had a response rate of at least 80%.
Source: AirDNA, Bakker et al. (2018b).

entire homes and apartments were offered during a certain number of days, and also how many were actually booked:

Table 1.2 Percentage of active entire homes available (booked) during a certain number of days in 2017

	Amsterdam		Berlin		London		Madrid		Paris	
<31	28.6	(56.3)	29.7	(67.3)	19.0	(58.0)	12.7	(47.1)	20.6	(60.5)
31–60	14.1	(14.2)	16.3	(10.1)	11.7	(12.3)	9.1	(11.0)	11.1	(11.4)
61–180	31.4	(24.0)	27.5	(15.9)	30.8	(23.7)	25.6	(25.5)	27.1	(19.6)
181–300	17.7	(5.0)	16.0	(5.8)	23.9	(5.6)	28.4	(14.2)	23.1	(7.1)
>300	8.2	(0.4)	10.5	(0.9)	14.5	(0.4)	24.3	(2.2)	18.1	(1.5)

Source: AirDNA, Bakker et al. (2018b).

In Madrid, London and Paris, four-fifths or more of entire house listings are available for more than a month. With the exception of Berlin, between 40% and 50% of entire homes are also actually rented out for more than 31 days. Properties offered for in excess of one month are less likely to be used for 'sharing'.

The most debated manifestation of the professional and commercial exploitation of Airbnb listings is hosts who do not just offer their own primary residence on the platform, but multiple units; these so-called 'multilisters' are usually understood to be commercial providers (Arias-Sans, 2015; Cox, 2015; New York State Attorney General, 2014; Slee, 2015). As we will discuss in Chapter 5, there are two different types of 'multilisters'. In the first place are those who buy or rent residential properties for the purpose of turning them into vacation rentals. A look at host profiles suggests that the typical 'multilister' of this type is a financially successful individual who invests his or her earnings – or previous Airbnb revenues – in residential real estate rather than in stocks or a savings account. In the second place are professional Airbnb intermediaries who may either represent home-owning residents or larger commercial parties active on the platform.

If we go back to our definitions, it is clear that the effects of investments to transform residential housing into tourist accommodation are contrary to the essential principles of 'sharing': they turn a scarce asset into idling capacity. For properties listed by management companies this may or may not be the case; however, as hosts, the activity of these intermediaries consists of traditional economic transactions, unrelated to any 'sharing'. As for the contact with locals, this may in practice remain limited to a professional representative employed by one of these companies, or by a real-estate investor driven by vacation rental return-on-investment.

Studies published by Airbnb itself typically claim that the vast majority – usually around 90% (Airbnb, 2012) – of hosts offer their primary residence on the website. This may be true, but the number of hosts is a mere

Table 1.3 Concentration of active listings, reservation days and yearly revenue in the hands of hosts with one or multiple listings, February 2017–February 2018

	Type of host	Number of hosts		Number of listings		Reserved days	Revenue
Amsterdam	1 listing	22,895	82.2%	22,895	57.3%	59.4%	59.4%
	2 listings	3,315	11.9%	6,630	16.6%	17.3%	16.2%
	3–10 listings	1,531	5.5%	6,068	15.2%	16.6%	16.6%
	More than 10 listings	112	0.4%	4,365	10.9%	6.7%	7.8%
	Total	27,853	100.0%	39,958	100.0%	100.0%	100.0%
Berlin	1 listing	33,356	78.9%	33,356	57.1%	56.1%	51.9%
	2 listings	6,220	14.7%	12,440	21.3%	20.2%	18.6%
	3–10 listings	2,573	6.1%	9,720	16.6%	18.8%	21.4%
	More than 10 listings	132	0.3%	2,918	5.0%	4.9%	8.1%
	Total	42,281	100.0%	58,434	100.0%	100.0%	100.0%
London	1 listing	63,203	73.1%	63,203	38.8%	33.3%	29.6%
	2 listings	13,457	15.6%	26,914	16.5%	15.9%	13.5%
	3–10 listings	8,601	9.9%	36,387	22.3%	24.7%	23.4%
	More than 10 listings	1,255	1.5%	36,585	22.4%	26.0%	33.5%
	Total	86,516	100.0%	163,089	100.0%	100.0%	100.0%
Madrid	1 listing	14,217	71.4%	14,217	36.9%	32.3%	28.8%
	2 listings	3,024	15.2%	6,048	15.7%	15.0%	12.3%
	3–10 listings	2,355	11.8%	10,063	26.1%	30.5%	29.1%
	More than 10 listings	311	1.6%	8,187	21.3%	22.2%	29.8%
	Total	19,907	100.0%	38,515	100.0%	100.0%	100.0%
Paris	1 listing	84,491	84.2%	84,491	62.7%	61.2%	53.7%
	2 listings	11,834	11.8%	23,668	17.6%	15.8%	14.2%
	3–10 listings	3,730	3.7%	14,222	10.6%	12.0%	13.3%
	More than 10 listings	325	0.3%	12,424	9.2%	10.9%	18.8%
	Total	100,380	100.0%	134,805	100.0%	100.0%	100.0%

Source: AirDNA, Bakker *et al.* (2018b).

smokescreen in the discussion about the concentration of properties in a few hands. In other words, it makes sense that the big 'multilisters' are a small percentage of all hosts, but the relevant number is the share of properties that they offer. Obviously, this 10% of 'multilisting' hosts represents far more than 10% of the offer. In London and Madrid, Table 1.3 shows that less than 30% of hosts are responsible for more than 60% of Airbnb listings.

As we will see in Chapter 5, these numbers also show the results of professionalisation: 'multilisters' are more successful on Airbnb – both in terms of reserved nights and of revenue – than regular hosts with a single listing. With 60% of the offer, the London and Madrid 'multilisters' make more than 70% of the total yearly revenues. The 1.5% biggest 'multilisters' in London are responsible for more than one-third of the total yearly revenues. That is, these 1255 hosts made on average €218,000 each in one year.

'Sharing' as a Push for Deregulation

With these analyses, we have entered an area where companies such as Airbnb have been extremely reluctant to provide any kind of 'sharing': data on the platform's performance and development. In the introduction we have discussed the need to use 'scraped' data as well as their limitations. The lack of data is an issue that concerns local governments as well as researchers. Unlike with traditional accommodation providers, no obligation can be imposed to disclose performance or visitor data. This makes it impossible to enforce regulations – such as taxation or rental limitations – but also limits insight into visitor streams. Agreements between cities and Airbnb, such as the one in Amsterdam, rely on information provided and on self-regulation adopted by the company. Both Amsterdam and the UK Office for National Statistics recently announced new monitoring methods to breach Airbnb's data monopoly (Kraniotis, 2016; Ward, 2016), but success in such a data 'arms race' seems not to be guaranteed.

The purpose of withholding visitor data is not just to dodge regulatory measures; they are an obvious source of competitive advantage in areas such as revenue optimisation or customer satisfaction (Garcés, 2018; McAfee & Brynjolfsson, 2012; Norton, 2015; Turnbull & Heinze, 2015; Xiang et al., 2015). It will be interesting for future research to monitor to what extent platform-based businesses such as Airbnb and OTAs will outperform traditional hotel companies in this area.

Although the underlying data themselves are not disclosed, Airbnb has developed an active policy of publishing impact reports produced by renowned consultants and based on their secret data. Not surprisingly, these reports underscore the economic impact of Airbnb visitors, the spreading of tourism to peripheral neighbourhoods and the contribution to the livelihood of lower income or non-traditionally employed residents. These findings, however, cannot be corroborated by independent studies. But in the absence of transparency on the data or methods used, the outcomes make little sense, such as the claim that almost all guests (93% in Amsterdam and Paris, 96% in Barcelona) want to 'live like a local', if we do not know what they have been asked, and how this concept has been operationalised.

The presentation of findings may also obfuscate the nature of the business, especially in the proportion of 'multilisters' on the platform. If we consider that the problem is that an important part of the Airbnb offer is concentrated in a few hands, obviously what is interesting is the percentage of properties offered by the different categories of hosts. It would be absurd to state that the concentration is limited because the hands in which it is concentrated are so few. Nevertheless, this is precisely the reasoning of the Airbnb reports when they proudly claim that the vast majority of Airbnb hosts (e.g. 87% for Amsterdam, 90% for New York) are sharers of their primary residences (Airbnb, 2012, 2013a, 2013b, 2013c, 2013d, 2013e, 2014a, 2014b, 2014c, 2015a).

These publications indicate that, besides hindering regulation and achieving a competitive advantage, the lack of transparency serves a third purpose: a marketing and lobbying strategy to attract guests and hosts and to convince city authorities to adopt benevolent policies. To support this message, strong lobbying investments have been reported for companies such as Airbnb as well as Uber (Guttentag, 2015; Lehr, 2015; Sottek, 2014; Sundararajan, 2016; Vekshin, 2015). Apparently the goal is to seek agreements such as with San Francisco, Portland, Amsterdam and Paris (Davies & Mishkin, 2014; France 24, 2015), which allow for further expansion in exchange for certain self-regulatory measures and the payment of tourist tax – the amount of which seems to be established by Airbnb itself.

As Slee convincingly argues in his book *What's Yours is Mine: Against the Sharing Economy* (2015), the 'sharing' narrative conceals a struggle of market forces against governmental regulation. 'It is not about building an alternative to a corporate-driven market economy, it's about extending the deregulated free market into new areas of our lives' (Slee, 2015: 24). Its secrecy has obscured the role of professional operators and venture capitalists backing the platform, and has allowed it to promote a rosy image of ordinary people opposing big business, with their emancipatory initiatives obstructed by obsolete bureaucratic rules.

Taxation, consumer protection, safety laws and social housing are some of the hardly controversial areas where governmental regulations used to restrain free market forces. In addition to the misleading 'sharing' belief, the global nature of this business has prevented local authorities from successfully intervening in its operations. Privacy protection has been the main argument for Airbnb not to disclose the data of its business operators. Eventually, whenever cities detect flagrant violations of their laws, individual operators are usually held responsible rather than the platform.

The struggle against regulations is, however, not just passive or elusive, but also active. The most overt and well-known intervention of Airbnb in the local debate was in San Francisco, where Proposition F sought to limit short-term rentals to 75 days per year and to ban short-term rentals of suites or apartments attached to homes. Whereas campaigners for affordable housing and other groups backing the proposal raised US$486,000, Airbnb invested over US$8m in the counter-campaign. The platform managed to spin the rejection of the measure in the referendum as a defeat of big business' interests:

> This election was a victory for the middle class. The Airbnb San Francisco home sharing community became a movement, showing up at the polls in large numbers and voting overwhelmingly against an effort designed by the hotel industry that targeted the right of the middle class to use home sharing as an economic lifeline. (Christopher Nulty, quoted by Booth & Kiss, 2015)

There are clear indications that Airbnb has also mobilised a more covert force of lobbyists to influence political decision makers at a local level, similar to what has been reported for Uber (Sottek, 2014). Benevolent regulations and tailor-made tax agreements were negotiated by the platform in cities such as Paris and Amsterdam. Prior to the 2018 Dutch municipal elections, Airbnb's European Public Affairs Manager visited the country's main cities and stakeholders in a campaign supported by professional lobbying agency Wallaart & Kusse Public Affairs. In Reykjavik, hotel owners accused politicians and prominent journalists of being active on the platform. With – often undisclosed – revenues per property in popular neighbourhoods of over US$10,000 per year, it is not unthinkable that conflicts of interests have benefited the platform in discussions about local regulations.

Extraction

So far, we have questioned Airbnb's inclusion in the 'sharing economy', as it neither intensifies the use of idle assets nor seems to strengthen community bonds. On the contrary, the platform encourages city residents as well as professional investors to transform housing into tourist accommodation, exploiting the price difference between the two. Fast global expansion has only been made possible by externalising financial and other risks to participating operators or 'hosts'. The platform organises market access to hosts and guests by charging both a commission. Airbnb is therefore an example, in Bauwens' definition quoted above, of 'a more parasitical form of capitalism, in which capital no longer organizes production, but facilitates P2P exchanges' (Araya, 2015).

The extraction takes place both directly and through a more profound economic transformation. In the first place, the platform benefits from the attractiveness of tourist destinations to which it has made no contribution: it has neither made a financial investment in the infrastructure or services required for tourist visits, nor has it contributed to the immaterial aspects of their attractiveness, such as local character and atmosphere. Its clients use these material resources and – through the phenomenon of 'overtourism' – contribute to the depletion of a city's immaterial resources, a development that we will describe as 'touristification'. To make things worse, its parasitical nature is underscored by the systematic evasion of tourist, income and corporate taxes to the destinations that are affected.

In the second place, urban vacation rentals have an impact on local economies. By creating additional demand for residential properties, housing prices can logically be expected to rise. This effect contrasts, however, with the assertion by the company itself that, for example, 'Airbnb helps Barcelona families earn extra money to stay afloat during these difficult times by opening their homes to visitors' (Airbnb, 2013e). It does not make any less sense to invert this reasoning: rather than enabling residents to afford their monthly rent, the platform is responsible for an inflation with which some residents only can cope if they participate in vacation rentals. In other words, Airbnb activity ceases to be voluntary as it extracts the inflated value of residential housing. The final consequence of this development would be the dystopian scenario in which city neighbourhoods are turned into privately owned timeshares (Oskam & Boswijk, 2016).

Summary: Neither Underutilised Assets nor Human Connection

In conclusion, the vast majority of the practices of Airbnb and other urban vacation rental platforms do not comply with Botsman's definition of a 'more efficient use of underutilized assets', nor with their own definition of 'commerce with the promise of human connection'. Instead, they imply the transformation of a scarce asset such as housing into tourist accommodation to increase its profitability. In this process, platforms such as Airbnb extract part of the increased value of the repurposed housing stock. As global businesses, they enter in conflict with local attempts to establish housing policies and to regulate commercial activities. Their lack of transparency is instrumental to this strategy.

2 Platforms or Two-sided Markets

> In the modern age, having icons on the homepage is the most valuable real estate in the world, and trust is the most important asset. If you have that, you've a license to print money until someone pushes you out of the way.
> Goodwin, 2015

Tom Goodwin is also the author of the famous quote: 'Uber, the world's largest taxi company, owns no vehicles. Facebook, the world's most popular media owner, creates no content. Alibaba, the most valuable retailer, has no inventory. And Airbnb, the world's largest accommodation provider, owns no real estate. Something interesting is happening' (Goodwin, 2015). This quote has been repeated in many business PowerPoints, sometimes to convey the message that today's world has lost interest in material goods, and merely seeks access. Of course, this is only partially true: if there were no solid base of material assets underneath these platforms they would not have gained any customer base. To be more precise, these platforms play a matchmaking role between buyers and sellers.

Two-sided Markets

This matchmaking role has been well documented in literature about platforms connecting two different customer bases or 'two-sided markets' (Caillaud & Jullien, 2003; Eisenmann et al., 2006; Evans & Schmalensee, 2016; Rochet & Tirole, 2004, 2006). Although it is true that the internet has reshaped and globalised the scope of this intermediary role, two-sided markets existed prior to the emergence of modern communication networks. To give a simple example, imagine a nightclub that owes its popularity to its matchmaking capacity and reputation: it must ensure that it attracts not only partner seekers but also potential partners. In case of a predominantly heterosexual audience, this means that visitors must be male and female in similar proportions.

A defining feature of two-sided platforms is the non-neutrality of their price structure: the distribution of the total price over the two market sides is designed to induce both groups of customers to join the platform, and thus to increase the volume of transactions (OECD, 2009; Rochet & Tirole, 2004). A greater presence of one customer group – in the previous traditional example, female guests – will increase the matchmaking network's value for the other group. Such network externalities lead to discriminatory pricing strategies with the purpose of expanding one customer base in order to add value for the other; the example of a free entrance on 'Lady's Night' illustrates how one customer base may subsidise the other side. There are a number of other examples that demonstrate and explain different distributions.

Newspapers, for instance, depend on readers and advertisers. Whereas the product is differentiated on the readers' side, it is less so for the advertisers: the latter need multiple platforms ('multi-homing') to reach all the readers who mostly choose to read one newspaper ('single-homing'). As a consequence, the newspaper will seek a large reader base, charging subscribers less than cost, while making its profits from advertisers who are obliged to multi-home (Armstrong & Wright, 2007). But also in the case of monopolistic platforms such as Yellow Pages, the benefits for the advertiser of each reader using the platform can be sufficiently high as to reduce the reader's price to zero (Armstrong, 2006).

The appeal of a game console increases with the number of available games; similarly, a game can achieve higher sales if a console has many users. In this case, the console buyers' side is subsidised by royalties and fees charged to (multi-homing) game developers; nevertheless, the opposite scheme was used for operating systems, which were paid for by consumers with software developers as 'loss leaders', leading to a dominant position for Microsoft (Rochet & Tirole, 2003). The difference in pricing strategies can be explained by the different price sensitivity of – mainly young – gamers, and of – often professional – computer users (Eisenmann et al., 2006; Heckman, 2016).

Credit cards serve the markets of payment makers and receivers. Cardholders are subsidised by the beneficiaries of payments, even up to the point of negative pricing through card benefits. On both sides we usually find multi-homers – cardholders with multiple credit cards and stores that accept different types of cards – although card acceptance policies can be used to steer store clients towards the least expensive option. Debit card holders, however, are not subsidised, which is explained by Rochet and Tirole (2003) by the fact that payment makers do not have to be wooed but use cards already in their possession. A final example is that of the supermarket, for which the normal model is to 'set retail prices to their consumers and to make take-it-or-leave-it offers to buy from their suppliers' (Armstrong, 2006: 684), thus using their bargaining power to optimise their customer base.

Winner-takes-all Competition

In summary, a nightclub with more potential female partners will attract more heterosexual male partner seekers; the more games there are available, the more gamers will buy the corresponding console; the wider the acceptance of a credit card, the more interesting it becomes to users, and vice-versa. These mutually enforcing network effects endow two-sided markets with increasing returns to scale, unlike traditional business models, where margins tend to decrease once their customer base has reached a certain size. This mechanism drives the emergence of oligopolistic or monopolistic platforms:

> Platform leaders can leverage their higher margins to invest more in R&D or lower their prices, driving out weaker rivals. As a result, mature two-sided network industries are usually dominated by a handful of large platforms, as is the case in the credit card industry. In extreme situations, such as PC operating systems, a single company emerges as the winner, taking almost all of the market. (Eisenmann et al., 2006: 3)

Insofar as the platform's exclusive task is to connect both customer sides, it will follow the logic of 'the bigger the better'; eventually, this will lead to a 'winner-takes-all' competition, in which the dominant customer base will continue to increase the platform's value to the customer base on the other side. Eisenmann et al. (2006: 7) analyse the conditions under which this 'winner-takes-all' competition occurs:

(1) 'Multi-homing costs are high for at least one user side.' It may, for instance, be too expensive or inconvenient to use two operating systems on a PC, or to buy two different gaming consoles.
(2) 'Network effects are positive and strong – at least for the users on the side of the network with high multi-homing costs.' The effect is that, considering that switching costs may be high, a user will critically look at the size of the market on the other side, e.g. the availability of software for the operating system, or games for the selected console.
(3) 'Neither side's users have a strong preference for special features' or, in other words, the connecting function of the platform prevails. As a consequence, the larger collection of software available for the Windows operating system was preferred, by many customers, to additional features offered by Apple Mackintosh.

Although the network effects are strong in urban vacation rental platforms, dominance in fully digital platforms can be more ephemeral than previously as switching costs are minimised: people can combine a range of social network apps, alternately use Lyft and Uber, and easily give up their iTunes collection to switch to Spotify (Tucker, 2018).

Platforms do not evolve as traditional monopolies, as their market power does not allow them to unilaterally set prices: excessive pricing will spark a negative feedback loop which will erode the platform's value to both market sides (Evans, 2009; McFarland, 2017). The discussion is whether the concentration of power in the incumbent platform may affect consumer rights, deter new entrants and thus block innovation, making it desirable to regulate markets. Regulatory actions against mobile telephony providers or against Microsoft illustrate this concern. A recent ruling about Amex forbidding merchants to steer customers towards using other payment methods – a policy analogous to OTAs' 'rate parity' – found this not to be a violation of competition rules as the two-sided dynamics made the policy not exclusively detrimental for merchants (McFarland, 2017). In other words, the debate is when or to what extent platforms display illegitimate anti-competitive behaviour, or if their innovations generate value by creating new synergies and efficacies (Bostoen, 2018; Garcés, 2018).

Online Travel Agents

As we argued in a scenario study on the future of online travel agents (OTAs) (Oskam & Zandberg, 2016), the evolution of electronic hotel distribution follows two-sided market dynamics. Expedia and Booking (formerly Priceline) act as matchmakers between hotels and travellers. While hotels need to 'multi-home', the completeness of the offer must convince hotel bookers to 'single-home', as the simplicity of 'all in one place' is the central proposition of the OTA. Their oligopolistic market power has not increased consumer prices – the subsidised market side – but has put hotel companies' margins under pressure.

The early interpretation and expectations of internet distribution also show how these dynamics – prior to the insights of Rochet and Tirole's publications in particular – were easily misunderstood. Porter (2001) discussed the influence of the internet on strategy, pointing at disintermediation and the increased bargaining power of consumers. His understanding that the internet is an 'enabling technology' rather than a disruptive development seems to be – from today's perspective – an underestimation. Nevertheless, just after the internet bubble burst, the idea that the internet was to some extent 'hyped' was more general, as is also observed by O'Connor and Frew (2002); this results in the forecast that the multiple channels existing at that moment would evolve but continue to co-exist, with the most promising growth to be found in the hotel chains' central reservation systems. Buhalis and Licata (2002) correctly predicted reintermediation, but they were also still sceptical about the business model and competitiveness of the new 'eMediaries'. Carroll and Siguaw (2003) drew attention to the disadvantages of the role played by intermediaries, identified as loss of revenue, diminished pricing control, and commoditisation.

Some OTA practices, such as 'rate parity' – equivalent to credit card companies discouraging stores to pass on card charges to customers – have become considered as abusive, and several countries have seen competition proceedings against Booking (Bostoen, 2018). The main threat to the OTAs' dominant position is what Eisenmann *et al.* (2006) qualify as 'envelopment', in which an adjacent platform provider incorporates the matchmaking functionality of the 'enveloped' provider into a multiplatform bundle. The envelopment would occur in a scenario in which the current symbiosis between Google and Booking – its largest advertiser – makes way for a hotel booking functionality in the search engine itself (Oskam & Zandberg, 2016).

Urban Vacation Rental Platforms

Urban vacation rental platforms do not own the accommodation they offer; they are pure matchmaking agents that allow accommodation seekers ('guests') and accommodation providers ('hosts') to find each other, with additional services facilitating the transaction. Payments are usually distributed over both sides, with a general preference for subsidising providers as offering the largest choice has become the primary differentiating feature. Both sides are likely to 'multi-home' as there are usually no entrance fees, even though the economic value of reviews can imply a switching cost for either side.

Urban vacation rentals are also a prominent example of a 'winner-takes-all' competition, with Airbnb as the undisputed market leader. The platform has achieved this 'incumbent' position as an early entrant with a more compelling proposition to the 'host' side than – at that time – exclusively subscription-based competitor HomeAway. To make its peer-to-peer (P2P) model work, Airbnb had to address two further issues in addition to getting hosts and guests on board: avoiding a direct negotiation and establishing trust as a condition for transactions to take place. As indicated by Guttentag (2015), marketing power is what set Airbnb apart from the traditional vacation rental market. The 'sharing' philosophy and the image of a warm and authentic community – as transmitted mainly in video testimonials – has been essential in persuading hosts and guests to join the network (Stern, 2010; Yannopoulou et al., 2013). But at the same time, direct transactions between the two market sides had to be prevented as this would lead to a one-sided business (Rochet & Tirole, 2004). The company does this directly through an algorithm that blocks messages containing phone numbers or email addresses, as well as by offering services to facilitate transactions such as credit card payment, pricing tools and insurance (Consigli et al., 2012; Hill, 2015).

Competing vacation rental platforms are scarcely differentiated in their offer. Pricing structures are similar as well, at least if we consider that platforms that subsidise 'guests' instead of 'hosts' are far less successful. Some platforms may originally have been stronger in traditional tourist areas – e.g. coastal vacation resorts – but with competitive pricing structures it would make little sense for 'hosts' not to 'multi-home', offering their property on different channels. Differentiation emerges, however, because of the counter-strategies of companies threatened by the platform disruption: OTAs and hotel companies. HomeAway was acquired by Expedia, onefinestay by Accor. The latter company is atypical because, despite the fact that it describes itself as the *'unhotel'*, it is more similar to a hotel than regular P2P platforms. Unlike the decentralised responsibility in other platforms, onefinestay guarantees service consistency through a 'gatekeeper' function – selecting listed properties – and by offering typical hotel amenities and services.

In practice, this will mean that a host who wishes to advertise an urban vacation rental will in the first place be attracted to the platform with the largest guest base, which currently is Airbnb. This host will probably multi-home, i.e. offer the listing on different platforms with similar conditions, unless he or she opts for specific characteristics incompatible with other platforms, such as the additional services of onefinestay. Without a rate parity clause, the host is also free to set prices in such a way that guests are nudged to use the platform that offers most advantages to the host, such as lower commission rates. A traveller looking for suitable accommodation will, on the other hand, be inclined to start the booking process through the platform with the largest offer because, assuming that hosts multi-home, apartments offered on the smaller platforms can also be found on the incumbent platform. Only if an identical offer is offered under more advantageous conditions will a multi-homing guest consider making his or her booking through an alternative platform.

The urban vacation rental matchmaking market will thus become increasingly concentrated: competing platforms would have to simultaneously offer advantageous conditions to both hosts and guests, unless they are perceived as the better matchmaker between its demand and supply base – which could occur in the case of 'platform envelopment' as discussed below. This market control differs from traditional monopolies or oligopolies in that the monopolist will not directly raise market prices. It may aspire to increase intermediation fees or commissions; this would require a strategic interaction between suppliers or jeopardise its client bases. Such coordination is unlikely, however, as the markets of oligopolistic platforms overlap instead of being divided.

Prices are not set by the platforms, but by individual suppliers – the hosts. A study of Airbnb demand in Vienna has shown this demand to be price-inelastic; travellers' 'love of variety' makes them cherish supply characteristics other than price. The authors conclude that there is room for revenue increases via price hikes (Gunter & Önder, 2018). This conclusion seems to be confirmed by the observation that Airbnb average daily rates approximate those of hotels in different cities (Bakker *et al.*, 2018b).

Despite its incumbent position in America, Europe and numerous upcoming destinations, a market that Airbnb has not been able to dominate is China, with a relatively small offer of 80,000 listings limited to the primary cities (Bergman, 2017; Custer, 2016). Airbnb's rivals in this market are Tujia (and the platform mayi.com it acquired in 2016) and Xiaozhu. China's housing market is less suitable for proper P2P sharing, because capacity is adjusted to family sizes and because it is culturally less

Table 2.1 (Selected) urban vacation rental platforms

Platform	Since	Offer	Scope	Home base	Specialisation	Cost structure
HomeAway	2004; acquired by Expedia in 2015	2,000,000	Global	Austin, TX		5–12% guests; 8–10% hosts or flat US$399 annual subscription
Airbnb	2007	3,000,000	Global	San Francisco, CA		6–12% guests; 3% hosts
Onefinestay	2009; acquired by Accor in 2016	2,600	London, Paris, Rome, New York, Los Angeles and Miami	London	Upmarket properties; hotel-like service	0% guests; hosts receive a flat nightly fee of the price set by onefinestay
Wimdu	2011	300,000	Global	Berlin		16% guests; 3% hosts
9flats	2011	200,000	Global	Hamburg		0% guests; 12–15% hosts
Tujia	2011	460,000	China	China	Serviced apartment provider	
Xiaozhu	2012	140,000	China	Beijing		

Note: Data as found through platform websites and user comments, January 2018.

accepted to lend one's residence to a stranger (Xiang & Dolnicar, 2017). Houses offered as vacation rentals are therefore primarily sought in secondary residences bought as investment. The two-sided model is taking off much more slowly than elsewhere, partly turning both companies into serviced apartment providers in a one-sided market rather than into actual platforms (CEIBS, 2017). The payment structure – although not clearly advertised on the companies' websites – seems to rely on guests' payments to the platform, with the platform paying a fixed rent to property owners, instead of a 'two-sided' cost distribution.

Table 2.1 presents a selection[1] of urban vacation rental platforms with their – somewhat volatile – characteristics.

Booking.com: The Threat of Envelopment by Other Platforms

A major competitive threat to two-sided platforms comes from rivals that offer a similar service as part of a larger bundle: the example is that of mobile phones which have incorporated a range of hardware functionalities, potentially also 'enveloping' a credit card functionality. The stand-alone platform would have to adjust prices on the money-earning side or assemble a bundle of similar value to compete (Eisenmann et al., 2006). Such bundling is also supported by a concentration of commercial services in internet media houses, the so called 'Scary Five': Google, Facebook, Apple, eBay, Amazon. These multifunctional platforms attempt to establish a captive audience – for instance, by packaging software and hardware – and monopolise its internet access through what we could call the limited 'shelf space of a phone screen' (Oskam & Zandberg, 2016).

In fact, competitive forces have been developing in this direction in urban vacation rentals with OTA Booking entering the market. Booking's website does not distinguish traditional and P2P accommodation and indicates that it has 1.6 million providers listed. As we have seen previously, accommodation seekers are the loss leader for Booking, which means that this market side does not pay commission while providers are charged 15% of the rental fee.

The market entrance of a platform that mainly targets the 'guest' side has only become possible because the 'host' side has previously been developed by stand-alone platforms such as Airbnb. Booking offers its 'guests' the convenience of a search in which the difference between hotels and apartments is blurred, with zero entrance costs, whereas 'hosts' benefit from the larger customer base interested in either type of accommodation. To illustrate how Booking taps into the provider market mobilised by urban vacation rental platforms, it should be noted that the website explicitly encourages 'hosts' to 'multi-home'.

The Napster Scenario

Between 1999 and 2002, the innovative file-sharing company Napster saw its turbulent rise and fall. Undergraduate student Shawn Fanning had developed a programme which allowed users to search and download files hosted on the computers of network members; the main goal of the programme was to facilitate the free exchange of mp3 music files. At the height of its popularity the programme had 80 million users. The company faced fierce opposition and legal action from the five big record companies – EMI, Sony, Warner, Universal and BMG, backed by a number of artists – accusing the company of 'piracy' and of undermining the economic fundamentals of music culture itself. The music industry, at that moment, mainly sought to preserve a business model dependent on sales of non-reproducible copies (CDs and DVDs) through retailers (Green, 2002; Honigsberg, 2002).

Napster tried to adopt a subscription-based model which would transform the music exchange community into an – extractive – platform; it was also obliged to filter copyrighted material. But despite the fact that online distribution could potentially become lucrative for the traditional record labels, a compromise or any type of collaboration proved unattainable: 'The companies wanted the enemy crushed and humbled' (Honigsberg, 2002: 487). Court decisions forced Napster to go offline in 2001. A year later the company went bankrupt.

For contemporary readers it is evident that, even though Napster has not survived as a company or a platform, neither is the retail of physical copies still mainstream. The so-called 'universal jukebox' model had familiarised the market with unpackaged music sales – the ability to download individual songs rather than albums – while at the same time it allowed for exposure to unknown or independent artists. Napster was first succeeded by new platforms such as Gnutella and Kazaa; more recently, streaming services such as Spotify or iTunes have triumphed in both music and video distribution, returning growth to the traditional music industry through subscription models (Shaw, 2017).

The winner-takes-all competition leads to a baseline scenario – i.e. 'the fundamental future with no surprises' (Hines & Bishop, 2013: 37) – of market leader Airbnb continuing to increase its market share, eroding its competitors. But the Napster history shows a potential alternative course of events, in which this foreseeable evolution is disrupted by a competitive counter-attack by traditional providers, in particular through legal action; also a security incident could be an imaginable explanation for a 'sudden death' of the platform. However, the disappearance of a platform does not entail a suppression of the innovation and the benefits that were already embraced by the market. In other words, it is less likely that the eliminated platform will be replaced by traditional hotels than by businesses connected to industry interests: the Netflix and Spotify of tourism.

Blockchain

Currently, the main threat of disruption for matchmaking platforms seems to come from developments such as artificial intelligence (AI) and especially the future evolution of distributed ledger or 'blockchain' technology. Their potential impact is under debate and may or may not be to a certain extent 'hyped'; if we think back to academic analyses of the future of e-distribution from around the year 2000, a degree of scepticism can be expected until new business models bring the development to a tipping point. The characteristics of distributed ledger technologies can be summarised as follows:

(1) the distribution of infrastructure and redundancy of data storage;
(2) the automation and standardisation of transactions;
(3) the reintroduction of digital scarcity;
(4) the disintermediation of transactions, markets and exchanges. (Lim, 2018)

The key feature of disintermediation makes the matchmaking function of platforms redundant, especially in the establishment of trust and the facilitation of transactions. The San Francisco initiative BeeNest provides an initial example of how blockchain may shape the future of urban vacation rentals, as it denounces the user fee and data vulnerability of not truly decentralised platforms such as Airbnb. Instead, the new service promises the following advantages for home sharers:

- Can't be controlled by any single entity
- Can't be corrupted by malicious attacks
- Are fully transparent, trustworthy & immutable (unchangeable)
- Have fast transaction speeds & low fees to any internet enabled computer in the world
- Blockchains empower people – by the people, for the people! They eliminate the need for middlemen & data silos across thousands of platforms selling your personal details – Users are in control of their own information and transactions. (Bee Token, 2017)

The BeeNest proposition revolves around the payment-arbitration-reputation (PAR) protocol which decentralises the verification of user identities, the transfer of payments and, through 'juror pools', the arbitration of disputes.

For blockchain alternatives to become a competitive threat to existing platforms, there are two – interrelated – obstacles: market power and technology adoption delays. The user advantages of a decentralised platform are simply outweighed by the size and distribution of the market leaders' offer: unlike for financial services, consumer choices are mainly determined by the availability of physical and geographically located assets. The rate of technology adoption is hard to predict, and will mainly depend on developments on the provider side.

The innovation brought by blockchain has parallels with those of transmission control protocol/internet protocol (TCP/IP), as it changed communication over a fixed infrastructure to the distributed network transmission of data packages. Based on an analogy with the introduction of TCP/IP, Iansiti and Lakhani (2017) suggest that the adoption will progress slowly in four stages. In the first stage are 'low-novelty and low-coordination applications that create better, less costly, highly focused solutions' (Iansiti & Lakhani, 2017: 123): examples of these isolated applications were email for TCP/IP, and Bitcoin in the case of blockchain. During a second stage, multiple parties start adopting highly innovative solutions in coordination, as is happening currently with blockchain in the financial industry or in supply chain management. The third stage of 'substitution' is again low novelty, but needs a high level of coordination as it applies the new technology to an existing service; examples are online stores for TCP/IP, or multi-party cryptocurrency transactions. Finally, during the last stage the innovation potential of the technology is used to create new solutions and revolutionise business models. Voice over IP would be an example of this last stage of adoption.

It took 30 years for TCP/IP to go through these evolutionary steps. Blockchain, however, may be embraced sooner as the consumer access to new services is largely in place. The analysis of the adoption stages helps understand how and where the innovations may be introduced. Rather than being displaced by new competitors, it seems more likely that in the first two stages platforms will integrate solutions to improve their internal processes and to smooth transactions with external but adjacent providers: for instance, between urban vacation rental platforms, OTAs and reservation systems. Depending on the evolution of the regulation debate, the technology may also be used for the exchange of information with authorities.

BeeNest or similar initiatives could become examples of substituted vacation rental services, either in isolation, in connection with incumbent platforms or integrated into the processes of these existing providers. Finally, we may see the transformational power of the new technology in solutions that are yet to be created. They could emerge within existing platforms, or existing brands might be displaced through 'envelopment' by a generic transaction service. These innovative solutions might entail distributed and direct holiday rentals among multiple parties, or even something we might again accurately describe as 'home sharing'.

Summary: Oligopolistic Market Power

Urban vacation rental platforms are matchmaking services that facilitate communication and transactions between accommodation providers and travellers. Increasing advantages of scale cause a concentration of market power in these services, in which Airbnb has emerged as the undisputed market leader. Without any changes to the business environment, it will be impossible for rival platforms to achieve the size and geographical distribution of its double customer base of both providers and travellers. The main competitive threats are therefore: envelopment by a broader service provider such as an OTA, displacement by a legalised service after defeat in the struggle with incumbents and with authorities, or obsolescence of its services because of technological innovation.

Note

(1) For a more complete overview, see Hajibaba and Dolnicar (2017).

3 Living Like a Local! But How Do Locals Live?

> The key lesson from cults that can be applied to all communities is what I call the great cult paradox: people join cults not to conform but to become more individual. Most people think the opposite is true – that people join because they're psychologically flawed or socially inept. This is due to the media's portrayal of cults that are destructive organizations. Most members of cults and cult-like organizations join for the same reasons that you or I would join anything. […]
> People joining cults and cult-like organizations – even the Marines or a corporate cult – all said the same thing: It doesn't change you; it enables you to be more yourself. And this is because you feel 'safe enough' to express your unedited self.
> Atkin, quoted in Passiak, 2017

The concepts of 'human connection' and 'community' are central in Airbnb's communication. As we have seen in the first chapter, this is despite evidence to the contrary if we look at actual rental practices: in general, Airbnb guests are not couch surfers looking for new friends, but they stay in entire homes where the contact with the host is more similar to the interaction we might have with a hotel receptionist. So, what is it that turns their stay into 'living like a local'? Or, what does Airbnb actually mean if it promises 'local' experiences?

Authenticity

The idea of 'living like a local' suggests that locals and visitors behave differently. Moreover, the proposition offered by Airbnb implies that its users will have an opportunity that hotel guests do not have: to experience 'authentic' local life as different from the tourist experience of the hotel guest. This raises a few questions: what makes a certain way of life local or 'authentic', where can it be found and why does it stay out of reach of visitors in traditional accommodation?

Authenticity is intuitively a straightforward concept, and in fact it is frequently used in literature as 'the real thing' as opposed to objects or experiences that are somehow fabricated, manipulated or staged. In this simplest interpretation, the Venice situated in Italy is authentic while The Venetian in

Las Vegas is an inauthentic representation. However, especially when referred to consumer experiences, the word is no longer understood as a simple dichotomy – either genuine or fake – but as a scalable attribute with an array of different interpretations. Venice may be experienced by visitors interested in contemporary everyday life in Italy as 'inauthentic'. The Disney Corporation, on the other hand, may or may not be believed to pursue a strategy according to its authentic values, while it is considered as the very archetype of fakery by others. A combined illustration of these last two examples is that the perceived loss of authenticity in Venice has been described as its 'disneyfication' – an equivalent to what we will call 'touristification' in these pages.

Similarly, tourist experiences can be divided into those that actually penetrate and participate in local life and culture, and those that are deceived by inauthentic and staged enactments of local life. Obviously the 'authentic' ones are the ones sought after: 'Sightseers are motivated by a desire to see life as it is really lived, even to get in with the natives, and, at the same time, they are deprecated for always failing to achieve these goals. The term "tourist" is increasingly used as a derisive label for someone who seems content with his obviously inauthentic experiences' (MacCannell, 1973: 592). This leads to a distinction – adapted from Goffman (1959) – between 'front' where the enactment takes place in front of an audience, and 'back' where only performers are allowed: the quest for authenticity drives the tourist to open up ever more profound stage settings of the 'back', all of which are, however, to some extent adapted to the visitor. This continuum replaces the sharp divide between the fakery of tourist visits and the genuine awareness of local life reserved for the intellectual traveller (MacCannell, 1973).

Since these early essays, the concept of authenticity has been amply debated in sociology, marketing and specifically in tourism literature. In his analysis of different interpretations, Wang (1999) distinguishes in the first place 'objective' authenticity (the quality of an object being the original, rather than a reproduction) versus 'constructive' or 'symbolic' authenticity (the extent to which the experiences of the tourist match the stereotyped image of the visited Other). This type of authenticity is ideological as it usually takes a historical manifestation of a culture to be the authentic or 'unspoiled' essence of that culture: this is the case, for example, in tourist representations of precolonial Maori culture in New Zealand (Taylor, 2001). Seemingly ancient traditions are, however, not rooted in an ahistorical essence but invented at certain moments in time to justify a division of power (Hobsbawm & Ranger, 1983). As visitors seek to transcend the enacted 'front' and penetrate into the local 'back', tourist marketing can fabricate a 'traditional' version of a culture that subsequently can become considered as 'authentic' (Peterson, 2005).

Wang distinguishes 'existential' authenticity as a third type of authenticity. In this interpretation of authenticity, the focus shifts from Other to Self, thus linking the concept to self-actualisation. The authenticity of tourist experiences is then no longer a feature of the visited place but is defined by the ability of the visitor to reveal authentic features in him or herself: 'Tourism is thus regarded as a simpler, freer, more spontaneous, more authentic, or less serious, less utilitarian, and romantic, lifestyle which enables people to keep a distance from, or transcend, daily lives. The examples include camping, picnicking, campfires, mountaineering, walk-about, wilderness solitude, or adventures' (Wang, 1999: 360).

Gilmore and Pine (2007: 5), when they define authenticity as 'purchasing on the basis of conforming to self-image', seem to blend the different 'authenticities' by identifying existential authenticity as an antidote to pervasive commercial – either objective or constructive – fakery. In other words, what matters is not the genuineness of the products or services offered by a company, but whether its strategy is 'true-to-self': the Walt Disney Company and its commercial offer is authentic as long as they do not betray the children or family orientation that defines the company's heritage. The apparent assumption is that consumers will seek to benefit from the contagious authenticity of the 'true-to-self' companies.

Tourism scholars have suggested a relationship between the perceived authenticity of the Other and the existential authenticity of the Self. The knowledgeable and affluent tourist is able to distinguish authentic cultural or historical features of ever more competitive destinations, while at the same time looking for meaning by participating in ethical consumption or in liberating activities (Yeoman et al., 2007). For Hall (2007), existential authenticity is a general need of modern man, not dependent on the genuineness or fakery of objects or destinations but derived from the connectedness to and engagement with the world experienced by the individual. Kolar and Zabkar (2010) provide empirical proof for a positive relation between object-based and existential authenticity.

A Community of True Believers

An important key to understanding Airbnb's marketing and its claims to authenticity is the book *The Culting of Brands: When Customers Become True Believers* by Douglas Atkin (2004), who a decade later would emerge as the Global Head of Community of the platform. The book takes the example of religious cults – such as the early Christians, the Mormons or the Moonies – as a demonstration of how a commercial enterprise can build an expanding community of devout followers by offering them a sense of belonging and meaning. Apple, the car brand Saturn, BMW motorcycles and VW Beetle are analysed as examples of 'cult brands'.

A cult brand must not target the average customer, but those who feel different: 'Instead of trying not to alienate anyone, you must target the alienated and simultaneously separate your organization from the mainstream' (Atkin, 2004: 18). These alienated consumers will be looking for an environment where their feelings of difference will be considered a virtue, and where they can safely express their opinions and choices.

The devotion to the cult then stems from the opportunities for '*self-actualization* in a group of like-minded others who celebrate the individual for being himself' (Atkin, 2004: 6). In other words, the cult creates the circumstances for what we have defined previously as 'existential authenticity'.

The path for an organisation to achieve cult status is:

(1) *determine* your potential franchise's sense of difference;
(2) *declare* your own difference with doctrine and language;
(3) *demarcate* yourself from the outside world; and
(4) *demonise* the 'other' (Atkin, 2004: 20).

Having identified its niche or deviation from the mainstream, the cult brand will be able to build its community by focusing on the people side of the customers alienated by its competitors. In Atkin's analysis, the message of Jetblue was: 'Don't be an airline' (Atkin, 2004: 39). It managed to build a customer base by passenger-friendly communication and operations. Interaction among consumers subsequently creates a sense of 'belonging' and allows a brand community to grow. New members – carefully selected – are to be welcomed with positive attention: the Unification Church's so-called 'love-bomb'.

Besides the unsettling recruitment guidelines, this vision endows the commercial organisation with a responsibility traditionally reserved for social or religious organisations. 'Brands function as complete meaning systems. They are venues for the consumer (and employee) to publicly enact a distinctive set of beliefs and values' (Atkin, 2004: 97); they 'can become the loci for groups of individuals unified by shared values, interests and identity' (Atkin, 2004: 200). Businesses thus absorb the ideological manipulation through which social forces are used to justify a certain organisation of society; its function is narrowed down to the specific interest of growing a brand or increasing sales.

Let us turn to a later presentation by the same author to see how the conversion of a commercial brand into a social and political movement can be put into practice. In the context of the 2014 debate on the legalisation of tourist rentals in San Francisco, Atkin (2014) demonstrates how to mobilise a client community, turning it into a grass-roots movement – visually compared by the presenter to the popular support for the Obama campaign. Large numbers of customers – hosts and guests – were contacted to obtain their commitment to the Airbnb cause. The company follows an escalating approach: 'Lower the barrier to entry; Make incrementally harder …; … But more rewarding' (Atkin, 2014: 23). This results in a 'commitment curve', in which the company attracts a growing core of active supporters (Figure 3.1). In the San Francisco case, the following stages were identified in this evolving commitment: sign petition, tell your story, attend action meetup, tweet senator, story postcard, oped, senator visit, attend rally.

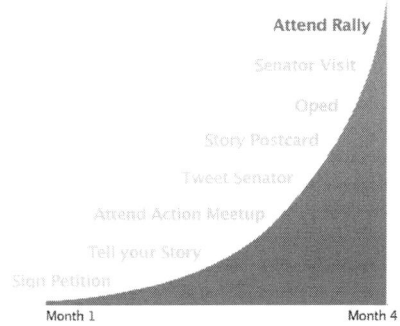

Figure 3.1 Atkin's commitment curve: Sign petition, tell your story, attend action meetup, tweet senator, story postcard, oped, senator visit, attend rally
Source: Douglas Atkin, Slideshare.

Evidence reproduced from the press illustrates the process:

> proponents demonstrated a much larger grassroots base. At hearing after hearing, hosts outnumbered opponents. [...][1] Supervisors continually ran into people who used Airbnb, liked Airbnb, and saw nothing wrong with Airbnb's practices. That's why you heard very little anti-Airbnb rhetoric from supervisors, including those who do not hesitate to castigate real estate speculators and Ellis evictors. (Atkin, 2014: 14)

And:

> I worked for a company that declared bankruptcy before paying me. But I knew I'd be safe financially because I had guests arriving. I'm grateful every single day because of Airbnb. (Atkin, 2014: 29)[2]

Atkin is also the co-founder of Peers, 'a grass-roots, member-driven organization that supports the Sharing Economy movement', whose activities and membership have been analysed by Slee (2015). A criterion to be considered part of the 'Sharing Economy' movement is, as Slee observes, not so much that an initiative gives shared access to a service, but rather that its distribution is technology driven. 'Out of Peers' 70 partners, over 60 are for-profit companies, and over 85% of the funding for Peers partners went to California companies' (Slee, 2015: 21). Slee concludes that Peers 'reflects this mix of communitarian intent and corporate self-interest', particularly in the interventions of Atkin who slips 'back and forth between the language of movement and the language of business' (Slee, 2015: 23), and that, therefore, 'the group functioned in part as a front for Silicon Valley lobbying' (Slee, 2015: 24).

In conclusion, the community building as envisioned by Atkin is related, at the level of the customer or the community member, to 'existential authenticity' in the sense that its purpose is the self-actualisation of members. But at the level of the business that promotes the community, it is a stratagem to increase brand value, sales and lobbying power. Disguising a for-profit business as a social movement is a 'constructive' inauthenticity as it exaggerates consumer benefits while silencing the actual interests behind the initiative.

Airbnb Promotion Campaigns

The different Airbnb promotion campaigns can shed a light on how the offer represents the concept of 'living like a local' and of authenticity in general. In the following pages we will analyse some of the promotional videos that are distributed through the company's YouTube channel.

Don't go to Paris

The principles of 'culting' become clearly visible in one of the most emblematic Airbnb campaigns. If a cult represents a deviation from the mainstream, but within acceptable boundaries for the alienated individual, it must first distinguish itself from the incumbent paradigm, and then demonise it (Van Burgeler, 2017). In this series of videos, the demonised 'enemy' is identified by visual cues and the voiceover as mass tourism: 'Don't go to Paris. Don't tour Paris. Please don't "do" Paris. Live in Paris.'

It is easy to imagine the discomfort and loss of individuality of the traditional tourist, taking selfies in front of the Eiffel Tower, riding a tourist bus or following the tourist herd on Segways (Figure 3.2).

Figure 3.2 Demonisation of mass tourism: 'Don't go there, live there.'
Source: YouTube.

The Airbnb message is 'Don't be a tourist', similar to the 'Don't be an airline' of Jetblue.

The alternative proposition is 'Live there'. Living, as the voiceover announces, equals doing 'your regular routine', but with a twist: the voiceover's 'Make your bed' is accompanied by children building a tent in a living room; 'Read a book' shows a man sleeping on the couch with a *Learn French* book on his chest (Figure 3.3). The 'normal routine' does not stand for what we usually do, but for what we would do if we were not held back by social constraints; that is, if only we were in contact with our authentic selves.

This 'existential authenticity' will allow us to better enjoy our time – hardly our destination – together with our travel companions: a partner and, in many cases, children. The connection with each other, often in a homely environment, is prioritised over curiosity about the city that is visited. This also means that locals play a minor role in these videos. They welcome us and show us the house; we also see them at dinner or during an excursion. However, the contact cannot be considered any more meaningful than the encounters a traditional tourist would have with tourist service providers.

The final shot shows a couple looking through their window at the Eiffel Tower, while the voiceover states: 'Wherever you go, don't go there. Live there, even if it's just for a night' (Figure 3.4). The scene can hardly be considered an unconventional tourist experience, and the shot might just as well have been taken from a traditional hotel room. The existentially authentic 'living' may bring the tourist couple closer together, but it in no way resembles how the vast majority of Paris residents live.

Figure 3.3 Existential authenticity: 'Read a book' the way a local would do it
Source: YouTube.

38 The Future of Airbnb and the 'Sharing Economy'

Figure 3.4 'Don't go there, live there. Even if it's just for one day.'
Source: YouTube.

Not yet trending

The 'Not yet trending' campaign is not as clear a demonisation of an opponent, although it emphatically differentiates the Airbnb proposition from the 'mainstream': that is, the 'already trending'. The videos show relatively unexplored destinations through the eyes of Airbnb hosts. This perspective is visualised through images where we see Airbnb properties and destination landmarks over the shoulder of dark silhouettes, shot from the back. Ironically – as also indicated by the word 'yet' – the goal and celebrated result is to send more tourists to these destinations. In general, these videos are therefore similar to destination promotion videos we get to see in contemporary TV shows. This apparent paradox can be solved if we understand that Airbnb is not about individual travel choices – 'Go where nobody else goes' – but about cult behaviour: 'Become part of our community by joining your peers.'

In Kanazawa, we follow local host Shungo. Shungo himself has fled Tokyo big city life to return to his town and to promote its traditional lifestyle. He and his wife host sake parties and dinners with traditionally prepared food for his non-Asian guests. His traditional robes, shopping at the market for food and artisanal production of food and sake support his undefined statement that 'Kanazawa is special because traditions are part of our everyday lives'.

In the 'Not yet trending' Azores, images of unspoiled nature and stereotypical villagers contrast with the eye-catching contemporary design of Airbnb properties. The destination does not target mainstream tourists as it does not 'sell the sun, the beaches and the parties'. But whereas our Japanese host represented traditional lifestyles, for the Azores the clues about how locals live are completely random and seem meaningless: 'Living inside a volcano means hot springs are a part of life' and 'When the fishing industry collapsed, hip-hop was born on the island'. But alongside these constructively authentic people, the visitor is depicted as a traditional tourist, relaxing and enjoying the natural beauty of the islands alone or with his or her travel companions.

In Puebla, food, cooking and markets are just as important as in Japan; colourful streets, mural paintings and partying are other shown aspects of local life. This is the only one of these three videos in which actual contact with locals is suggested visually and in textual cues, although the latter remain linked to the commercial service encounter between guest and host: 'I give a sense of home through my food. Guests always leave with a Mexican mom.'

Experience Hawaii and New York with locals

There are two 'Experience your destination with locals' videos that, in a similar way, promote 'off the beaten track' tourism, but to more traditional destinations. Local Airbnb hosts are the protagonists of these videos; tourists remain spectators. The activities promoted by the 'experience hosts' – as part of their commercial tourist businesses – derive their authenticity, in the Hawaiian video, from their roots in local and native traditions. The visitor groups observe, learn (Figure 3.5), and finally reach bliss, as expressed by laughing, applauding and bonding among the visitors and the tourist entrepreneur and among the visitors themselves. In other words, these videos link constructive authenticity to existential authenticity in a way that is similar to the reasoning of Pine and Gilmore.

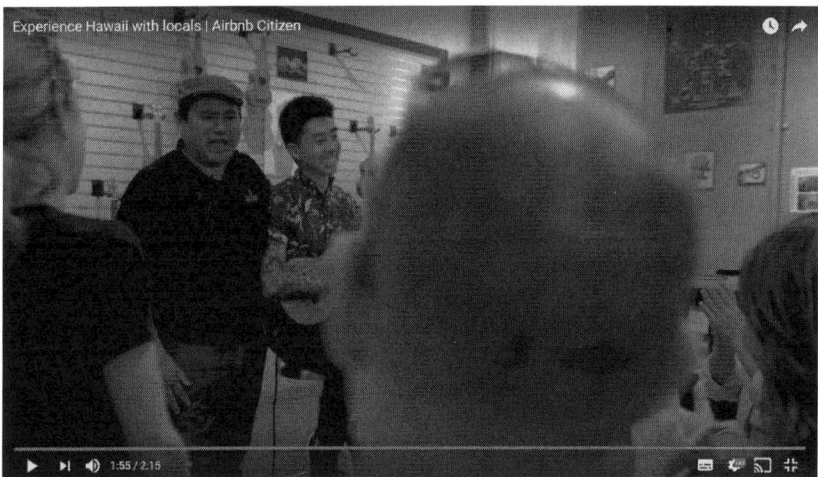

Figure 3.5 Observe and learn ...
Source: YouTube.

Sailing handcrafted canoes, playing the hula or building ukuleles are probably just as distant from Hawaiian daily lives as the Eiffel Tower is from the window of an average Paris resident. In this sense, Airbnb builds on the constructed images of tourist destinations. 'Living like a local' must not be understood as an attempt to integrate visitors into local communities, but rather to build small communities of visitor groups and specific local experience hosts, as a way for both to reveal their authentic selves.

The same can be said of New York. The daily lives of New Yorkers may be more tedious than the constructed vibrant image of the city, and authenticity can hardly be sought in the traditional lives of New York neighbourhoods. As an alternative, Airbnb's video takes us to different subcultures evenly spread over the boroughs: local artists in Brooklyn, a hipster urban farmer on Staten Island, street dancers in the Bronx, immigrant culture in Queens and contemporary artisans on Manhattan. These images of unknown borough life – intertwined with fast motion shots of landmark bridges and downtown traffic – are probably especially appealing to domestic tourists with a previous knowledge of tourist New York. Again, however, the purpose is not to get an in-depth understanding of local ways of life, but to bring out the best in yourself, as a tourist, through friendship with the individual local host.

Authenticity is Bliss

Airbnb videos depict the local as an exotic friend with whom we have more in common than we expect. This means that there is a delicate balance between familiarity and difference: most hosts are in the millennial or under-40 age groups, they are known by name and show us their world through their eyes. But a certain degree of exoticism is also required to link the local to the travel destination. Thus, along with the Portuguese hiphop artist we meet older farmers on the Azores and see locals drive a Renault 4. We can relate to the contemporary party crowd of Puebla but also to the Mexican mom serving local dishes. Social host Mai Lê takes us by the hand to discover the exotic street life of the Bronx.

The visitors play a secondary role in these videos: they are spectators like ourselves. Ethnically diverse, they in general pertain to the same age group as the host. The visitors mostly appear in small groups of around six people. The situations presented by the hosts invite the visitors to engage in unfamiliar activities. The discovery of 'authenticity' in these experiences establishes a bond with either the host or other group members, or both. This blissful discovery is the climax of each scene, celebrated with laughter, facial expressions of surprise and physical contact like kisses and high fives. The visitors have now been initiated in the local cult of authenticity (Figure 3.6).

Figure 3.6 ... to bond and reach bliss
Source: YouTube.

> ### Summary: Cult Admission
> The analysis of Airbnb campaigns shows that the promise of 'living like a local' should not be taken literally. Airbnb users are mainstream tourists, but liberated from the compelling scripts of mass travel. They seek experiences that are 'true to self' in what we have categorised as 'existential authenticity'. This allows them to enter the Airbnb community of 'true believers'. The admission ceremony, which we have witnessed in a number of videos, ends with a moment of bliss.

Notes

(1) The following sentence was omitted from the original: 'Hosts had financial motives and personal stories, while opponents were housing advocates focused on the negative implications on citywide housing affordability and availability' (Shaw, 2014).
(2) Article (Badia, 2013) displayed on the slide, incorrectly quoted in the verbal notes to the presentation.

4 Millennial or Mainstream: The Airbnb Guest

> But things are changing, and one of the reasons why is the digital natives, or Gen-Y. They're growing up sharing – files, video games, knowledge. It's second nature to them. So we, the millennials – I am just a millennial – are like foot soldiers, moving us from a culture of 'me' to a culture of 'we.' [...]
> I realized it was happening because of four key drivers. One, a renewed belief in the importance of community, and a very redefinition of what friend and neighbor really means. A torrent of peer-to-peer social networks and real-time technologies, fundamentally changing the way we behave. Three, pressing unresolved environmental concerns. And four, a global recession that has fundamentally shocked consumer behaviors. These four drivers are fusing together and creating the big shift – away from the 20th century, defined by hyper-consumption, towards the 21st century, defined by collaborative consumption.
>
> Botsman, 2010

As we have seen in the previous chapter, the promoted image of Airbnb guests is that of ethnically diverse millennials looking for connection with each other and especially achieving blissful self-actualisation through that connection. This is in line with the 'utopian spin' of the so-called sharing platforms. In addition to this strongly marketed profile of the Airbnb guest, a good understanding of who these users are and of what drives them is also hampered by the inevitable limitation of all Airbnb research. Since we do not know who they are nor where they are, how can we reach them to ask them questions and, even if we do, how can we evaluate the representativity of our sample?

With the use of expensive panels out of reach of many academic researchers, it is usual in Airbnb studies – and in studies of sharing in general – to resort to convenience samples, either in the social environment of the researcher or through Amazon Mechanical Turk. These studies should therefore be read as exploratory research and not as final evidence, despite the fact that some of them are methodically thorough and do provide interesting insights. It is not unthinkable, however, that a selection has occurred that reinforces a bias towards respondents inclined to support the utopian motivations for using sharing platforms.

In other words, sharing advocates such as Botsman (Botsman & Rogers, 2011) and Gansky (2010) have promoted the belief that sharing activities are environmentally sustainable, strengthen human connections and, by lowering costs, empower larger groups to have access to goods or services. These claims constituted the programme of a social movement and were not based on studies of the practical shape of and relations in the sharing economy. In the early days of the movement, it is evident that its followers or practitioners were attracted by these promising messages rather than by the sharing praxis as they could have observed it. The consequence would be that in earlier studies these followers of the movement are overrepresented, especially since sampling was not random and in some cases the denomination of 'sharing' was applied generically to all activities or platforms, including those for which this categorisation is questionable. It is like going to a rally of a political party and questioning people about their motivations to vote; the questionnaire will give insights into the political movement but will not lead to generalisable conclusions about how voters – users – outside the party are motivated.

Motivations to Participate in Sharing

Besides this partisan literature, the motives to use urban vacation rentals have been explored in consultancy reports, in reports issued by Airbnb itself and to a limited extent in academic literature. Commercial studies – including those published by the platform – combine the advantage of access to large user bases with the shortcomings of a lack of transparency regarding methods and a limited need to critically examine assumptions. It is important to realise that there is a general limitation to all motivation studies caused by the decentral development of rental platforms. Airbnb users in inexpensive destinations – e.g. in Eastern European cities – may be driven by different motives from visitors to London, New York or Amsterdam; the same is probably true for city holiday goers compared to more extended rentals of properties in more traditional tourist destinations, such as coastal towns and villages. Finally, the growth and acceptance of the phenomenon is likely to make a difference: the first air mattress crashers in 2008 probably had a different profile from the mainstream user of Airbnb in 2017.

Not surprisingly, Airbnb studies highlight those traits of its users that fit its image as a progressive counter-culture. These reports are made by renowned consultancy firms and support the platform's lobbying or marketing campaigns. A series of city reports underscore the positive economic, social and environmental effects of Airbnb tourism on destinations. In general, well over 90% of Airbnb guests respond that they are motivated to 'live like a local', and close to 90% are looking for 'off the beaten track' experiences (Airbnb, 2012, 2013a, 2013b, 2013e). Although the average age of users reported is 36 (Airbnb, 2013e), a later report with a stronger marketing focus claims that 60% of Airbnb guests are millennials (Airbnb, 2016a). In the case of Boston, 95% of guests were identified as families (Airbnb, 2014c); a later report published to indicate the appeal of the family market shows that 83% of families were driven by the desire to 'live like a local', 93% were motivated by having more amenities available than in a hotel and 96% chose Airbnb to save money (Airbnb, 2016b).

Also, commercial and commissioned reports have addressed the demographics and motivations of Airbnb users, with a special interest in the competitive effects for traditional businesses. In a study based on 383 online interviews, Olson (2012) finds a stronger appeal among millennials and particularly GenXers than among babyboomers, as well as among higher income groups as also reported by Tussyadiah (2015). This study detects rational benefits, mainly financial and environmental, as well as emotional benefits: generosity – 'I can help myself and others'; and community – 'I'm valued and belong'. A Morgan Stanley survey with over 4000 respondents in four countries (Nowak et al., 2015) confirms the financial motives of Airbnb users, with 'cheaper price' as the main reason for use for 55% of respondents, followed by location (33%) and authenticity (31%). Of tangible amenities, the availability of a kitchen was mentioned by 25%. The report concurs with Olson that users skew wealthier, with around 66% of US Airbnb users earning over 75,000 US$/year. EY España (2015) finds, among 2203 visitors to Spain from four different European countries, the age distribution, education level and travel attitudes of those who stay in urban vacation rentals to be very similar to those who stay in regulated accommodation. Again, price is mentioned as main reason of choice by 52.7%, followed by independence ('cooking, eating when I want', 45.8%), with the relation to the host (6.1%) qualified by the authors as a marginal motive.

A study based on an online survey of 1548 US millennial travellers (20–36 years, annual household income US$35,000 or higher; ResonanceCo, 2017) finds, for Airbnb, an 'overt alignment with US millennial traveller priorities', their main destination consideration factor being: safety (57%) – a critical factor increasingly managed by the platform; price (52%); and whether English is spoken (44%). The report highlights the 'Don't go to Paris' campaign we examined in the previous chapter: 'Of course it's another direct hit with our respondents, who overwhelmingly crave personal, local and immersive experiences in their destinations. Venturing out of their comfort zones and Learning New Things are prioritized by 85% of our respondents.' Curiously, it also states that millennial Airbnb users are influenced by the 'deep background briefings' by fellow travellers, provided *sans* marketese' (ResonanceCo, 2017: 56).

Despite these empirical findings, other consultancy reports with a more conceptual approach reproduce characterisations from partisan literature, thus contributing to their clients' competitive strategies – as well as the more general image of urban vacation rentals – being distorted by the 'utopian spin'. A different division of EY states (Roth & Fishbin, 2015) that Airbnb caters to the need for authentic and unique experiences. Deloitte (Langford *et al.*, 2017) concurs with this view and compares the drivers behind Airbnb to factors driving the popularity of craft beer, locally sourced and organic food, or 'indie' music. In the Future Foundation's *Amadeus Future Traveller Tribes 2030* report (Chiarelli & Warriner, 2015), the use of 'sharing' accommodations is linked to the personas of *cultural purists*, *social capital seekers* and *ethical travellers*, in the latter case explicitly as a protest against the established order: 'One dimension of 2030's consumer ethics may be anti-corporate, anti-globalisation or anti-urban' (Chiarelli & Warriner, 2015: 64).

As discussed, academic studies derive their main limitation from the fact that their sampling method may reinforce a predetermined focus. Hamari *et al.* (2016) have studied the motivation to participate in sharing activities with a survey among 168 users who had reacted to a call in the Sharetribe newsletter. These authors distinguish intrinsic and extrinsic motivations: an ideological inclination towards sustainability and enjoyment, on the one hand, versus reputation and economic benefits, on the other. Those in favour of sustainable solutions have a more positive attitude towards collaborative consumption, but this is not necessarily reflected in their actual sharing behaviour. Enjoyment is correlated with both attitude and behaviour, and economic benefits also have an effect on behaviour.

The study of motivations to use peer-to-peer (P2P) accommodation, specifically, by Tussyadiah (2015), is based on a questionnaire distributed through Amazon Mechanical Turk ($N = 799$). This study finds that users of accommodation sharing are better educated, have higher incomes and are more assiduous travellers than non-users; users also differ from non-users in their personal travel innovativeness trait, while both groups are similarly open to technological innovations. Main drivers for the use of P2P accommodation are sustainability considerations, community (i.e. interacting with locals) and economic benefits; lack of efficiency – i.e. unfamiliarity with the phenomenon or with the platforms – in the first place, and trust issues and lack of economic benefits withhold people from participating.

This study was replicated in Finland with 1246 respondents contacted through the M3 Online Panel (Tussyadiah & Pesonen, 2016). Except for gender and age – Finnish users were mainly male with a stronger presence of younger age groups – the demographics of users are similar. Also the drivers for use are similar, with female Finnish users more attracted by the economic benefits and older Finnish users more disposed to use the platforms because of social appeal (sustainability and community). Whereas in the United States lack of efficacy – unfamiliarity – is a stronger barrier to use for less well educated respondents, in Finland older users are less aware of P2P accommodation. The other barrier for Finnish travellers was (a lack of) value, which combines the findings of trust and lack of economic benefits in the American study.

In a first conceptual article, Guttentag (2015) analyses Airbnb as a disruptive innovation in line with Schumpeter's (1942) concept of 'creative destruction'. The innovation underperforms with regard to prevailing products' key performance attribute(s), but offers a distinct set of benefits to the market, in terms of cost, convenience or ease of use. This means that such a disruption initially appeals to the lower end of the market, and gradually incorporates more features to become mainstream. This article suggests that relatively low prices may be a first driver for adopters, combined with other benefits such as the feeling of being in a home, local advice and practical residential amenities. Finally, the 'living like a local' aspect and the possibility of staying in a 'non-touristy' area are mentioned, supported by quotes from the *Lonely Planet*'s Barcelona guide and *The New York Times* travel blog.

In subsequent work, the same author (Guttentag, 2016; Guttentag *et al.*, 2018) studies Airbnb user motivations, using a survey with (844) respondents recruited mainly through Facebook travel groups and Amazon Mechanical Turk, besides other channels, with the intention of reducing bias through this combination. Unlike previous studies, these analyses approach the Airbnb users as a heterogeneous group, with different motivations. The differently motivated users are clustered in five segments:

(1) *Money savers*. This group is younger than average (62.9% under 30 versus 53.2% total) and less likely to travel with children (3.3% versus 10.3%).
(2) *Home seekers*. Motivated by available space, household amenities and homely feel, this segment is older on average (23.7% over 41 years versus 16.8% average), better educated (35.4% hold a graduate or professional degree versus 29.7% average), travel with more companions (2.27 versus 1.79 average) and with a partner and children (respectively, 64.9% versus 57.6% and 22.3% versus 10.3%). They mainly stay in entire homes (92.0% versus 71.0%), for more nights (5.72 versus 4.24 average), and have been more frequent users of the platform (5.81 times versus 4.56 average).
(3) *Collaborative consumers*. The segment of users that is actually motivated by the principles of 'sharing' is slightly older than average (47.1% are 30 and under versus 53.2% average) and less affluent (27.3% believed their household financial status to be 'just below average' or lower versus 22.3% average). They travel more internationally (66.9% versus 60.2%), with fewer companions (1.26 versus 1.79) and use shared accommodation (55.2% versus 29.0%). They are more likely to be backpackers (25.0% versus 17.8%) and to have been Airbnb hosts themselves (14.5% versus 9.4%).
(4) *Pragmatic novelty seekers*. These users combine novelty with home benefits (as listed for the 'home seekers'). This segment is young (59.3% are 30 and under versus 53.2% average) and travel in groups (2.00

companions versus 1.79 average), using entire homes (90.1% versus 71.0%). They are unlikely to be backpackers (12.6% versus 17.8%) and have used Airbnb less frequently than average (3.71 times versus 4.56).
(5) *Interactive novelty seekers.* Motivated by novelty and interaction, this segment is looking for 'living like a local' and off the beaten track experiences. These users choose shared accommodations more often than average (47.4% versus 29.0%), for shorter stays (3.39 nights versus 4.24) and with fewer companions (1.50 versus 1.79). They are more likely than average to be backpackers (27.0% versus 17.8%) and have not used Airbnb often (3.43 times versus 4.56 average).

Because of the non-random sampling, the studies are not conclusive for the size of these five segments, but suggest that these are more or less similar (Guttentag, 2016). Therefore, on the basis of these studies a cautious conclusion can be drawn that only two-fifths of the Airbnb users actually correspond to the marketed image of alternative travellers. Besides, Airbnb attracts a younger crowd interested in economical travelling, and groups of people travelling in larger parties or with family who are especially drawn by the tangible advantages of Airbnb over hotels: square metres, practical amenities such as kitchens and washing machines and the independence that these features give the traveller. This is in line with Gunter and Önder's (2018) finding for Vienna that Airbnb to a certain extent provides a complement to the traditional hotel offer in some cities since Airbnb guests are apparently looking for a type of hotel room (e.g. suite rooms) which is currently in short supply

Later studies add little to the segmentation provided by Guttentag, although they contribute to the understanding of the generic motivations and attitudes of Airbnb users as a whole. A study among 294 South Korean students finds enjoyment and reputation to be significant antecedents of attitude towards Airbnb, rather than sustainability or economic benefits (Yang & Ahn, 2016). A model developed by Kam Fung So *et al.* (2018) uses an online panel (Qualtrics online sample, $N = 500$) to measure the effect of different motivations and constraints, with significant results determining the attitude towards Airbnb for price, enjoyment and home benefits on the one hand, and distrust and social influence on the other; enjoyment is the strongest predictor for both attitude and behavioural intentions. Finally, a survey among 1426 international travellers, recruited through social media, reports significantly higher scores for Airbnb users on the Big Five personality traits of openness, conscientiousness, extroversion and agreeableness (Pezenka *et al.*, 2017).

Guest Satisfaction

For a platform unable to guarantee a consistent service, guest satisfaction in Airbnb is mysteriously unanimous and high. The reciprocal review system – guests rate the Airbnb listing and hosts, but hosts also rate the guests – has been demonstrated to create an upward bias (Ert *et al.*, 2016; Zervas *et al.*, 2015), causing practically all ratings to be between 4 and 5: Fradkin *et al.* (2015) report 20% 4-star and 74% 5-star ratings (2015). The latter study shows with field experiments that this bias is not simply due, however, to review retaliation, but rather to review selection – guests with positive experiences are more likely to review – as well as to the awkwardness of leaving negative comments about hosts or guests after socially interacting with them. These attitudes are referred to, respectively, as strategic reciprocity, sorting and socially induced reciprocity.

Despite this bias, the absence of consumer protection makes the level of satisfaction improbable. Sites such as airbnbhell.com (*Uncensored Airbnb Stories from Hosts & Guests*), but also the 1 million US$ host property insurance introduced after serious incidents (Consigli *et al.*, 2012; Satama, 2014), are indications that deceptions and dissatisfaction also occur in the world of 'sharing'. However, an observation in many interviews with Airbnb guests is that they tend to blame themselves for disappointments; as this is part of the 'sharing' game, they are fully responsible for selecting the right offer. This phenomenon is possibly related to social reputation concerns in this market; considering the rental as a regular commercial transaction may affect the 'sharing' cool of the participant.

Studies of Airbnb guest satisfaction and loyalty seek to identify the characteristics of 'postmodern consumption behaviour', partly determined by different factors from satisfaction in traditional service encounters, but partly also by the usual utilitarian considerations. Lalicic and Weismayer (2017) show that satisfaction is strongly mediated by the perception of authenticity. Mao and Lyu (2017) find that the expectation of unique experiences has a positive effect on the attitude towards Airbnb, but that perceived value has an even stronger effect. A Norwegian master's thesis (Visjø & Slevigen, 2017) finds that consumer loyalty is determined by calculative commitment – price and cost savings – and by affective commitment, but not by sustainable commitment; emotions strongly predict satisfaction and affective commitment. The finding of Liang *et al.* (2018) that *transaction-based satisfaction* – i.e. the various aspects of making the purchase before the transaction is completed – has a considerably stronger effect on repurchase intention and trust than *experience-based satisfaction*, caused by the actual stay, further nuances the image of the traveller driven by different, less utilitarian, motives from the traditional consumer.

Airbnb for Business

Initially Airbnb mainly appealed to cost-conscious leisure travellers; Zervas *et al.* (2017) found Texan hotels that did not cater to the business segment to be most affected by the rise of Airbnb, with business travellers preferring hotels that offered work and conference facilities. Since then, the business use of Airbnb is reported to have seen sometimes spectacular growth, albeit based on low absolute numbers: according to a study by Certify, in 2016 the platform represented less than 1% of accommodation bookings in business travel (Sheivachman, 2017). Airbnb itself estimated 10% of its bookings to be for business rather than leisure (B.R., 2016). Morgan Stanley reported 12% of business travellers to have used the platform at least once in 2015, with a forecast of 16% for 2016, a more moderate growth than for leisure travellers: 33% versus 50% (Nowak et al., 2015).

In July 2016 Airbnb launched 'Airbnb for Business'; six months later the company tweeted that '50,000 employees of more than 5000 companies have booked with Airbnb for Business since its launch'.[1] It also segmented its offer by making a 'Business Ready' label available for listings with certain quality standards (a 4.8 rating minimum over the past year with at least five ratings), work amenities, safety features, self-check-in and a minimum response rate of 90% within 24 hours. Its growth strategy has led to partnerships with conference organisers (Oates, 2016) which, after all, was the origin of the company. The Global Business Travel Association reported that in the second half of 2016, business policies that accepted employees using platforms such as Airbnb for business travel had risen from 28% to 30% (Sampson, 2017). Despite all the buzz, the success of Airbnb for Business can be questioned. It may be hard to reconcile the marketed 'off the beaten track' image with the more mainstream requirements of business travellers. As a hypothesis, companies that centralise travel bookings are probably more reluctant to use the platform, since – as we have seen earlier – bad experiences are blamed on the booker rather than on the platform.

Further Research into Airbnb Guest Demographics

On the basis of the international literature, especially the segmentation study by Guttentag (2016), several students conducted brief exploratory studies to verify and validate these findings for Airbnb visitors to Amsterdam. A first study used host interviews to gain insight into guest profiles. The hosts declared that in general their users were professionals and experienced Airbnb users between the ages of 20 and 40; this was also partly due to the fact that guests with less desirable profiles – younger, or with a lack of affinity as expressed in written communication – would be less likely to be accepted. They were often groups of friends, couples, or families with children under the age of 12. The estimated length of stay was three days or a long weekend. When asked about the travel motives of their guest, all said they wanted to 'live like a local'; however, at the same time the most usual inquiries were about the traditional tourist attractions. The reason for their visit to Amsterdam was going to a special event – such as King's Day or the Gay Parade – visiting friends, cultural visits or bachelor parties (Van Loenen, 2016).

A second study quantified user motives using two different surveys. A first survey approached random sampling by asking every 10th visitor queuing in Amsterdam tourist hotspots (Anne Frank Huis, Rijksmuseum, Madame Tussaud's and Dam Square) for the characteristics of their stay ($N = 390$). Of this sample of leisure visitors, 28.7% were day visitors; among those who stayed overnight, 62.9% stayed in a hotel and 18.7% used Airbnb, while 12.9% used other accommodation. A second survey was sent to 400 Airbnb guests who had visited Amsterdam through concierge service Iambnb, obtaining 69 replies. The average age of these respondents was between 36 and 45; they were mostly family, friends or partners travelling in groups of 2.4 on average, and proceeded from North America or Europe. On average, they spent between €51 and €100 a day and had chosen their accommodation because of price, location or amenities, in this order. Twenty per cent of these Airbnb users indicated that they had not visited mainstream tourist attractions, while 80% responded that the location in the city centre and near touristic attractions was important.

This study found that the number of square metres was more important for older guests, guests who stayed longer and unsurprisingly for larger groups. Older users are less concerned about price; receiving insider information and interaction with the host or locals is important both for these older users and for guests who spend between €200 and €250 per day. Amenities were an important factor for families and for people who travel alone (Van den Bent, 2016).

In a survey distributed mostly through social media ($N = 238$) 64.8% of Airbnb users were likely or very likely to reuse the platform in the next 12 months and 66.8% would recommend Airbnb. Of 180 frequent Airbnb users, 72.8% indicated that they were satisfied or very satisfied with their recent stay. Low cost and convenient location were found to be the most important drivers for loyalty, with the 'local experience' being relatively less important (*somewhat agree*). Ideological advocacy of 'sharing' was not found to be a driver of loyalty (Kleinherenbrink, 2016).

A study of the competitive effects of Airbnb on hotel use was based on a survey ($N = 190$) distributed in mid-price, upscale and luxury (3-, 4- and 5-star) hotels. Guests of mid-price hotels were significantly more likely to alternately use hotels and Airbnb. Lower income groups (under US$20,000 yearly) and the income group between US$60,000 and US$75,000 per year were also more likely to use Airbnb alongside traditional accommodation. Luxury travellers were more likely to agree that a hotel was a better option than Airbnb in comparison to upscale and mid-price travellers. As indicated in open questions, reasons for hotel guests to choose Airbnb were categorised as: 'more space, personal contact, convenient location, cheaper, comfort of house, local experiences and house facilities'. Reasons to choose a hotel rather than Airbnb were: 'prefer hotel, unfamiliar with Airbnb, safety of hotel, lack hotel service, length of stay, more expensive and not on hotel booking platform'. For mid-price travellers, personal contact was a more important reason to switch to Airbnb. For shorter stays, in general, hotels were found to be a preferable option over Airbnb (Visser, 2017).

With a quota sample equally distributed over Guttentag's five motivational segments, a study based on street surveys ($N = 150$) explored the guest satisfaction of Airbnb guests with their accommodation and with Amsterdam. Overall satisfaction with the accommodation and with the destination was high. The 'home seeker' segment showed the highest satisfaction with their accommodation, and the 'money saver' the lowest. 'Home seekers' showed significant differences from some of the other segments on satisfaction with space and household appliances. 'Collaborative consumers' differed in their satisfaction concerning 'contact with locals', while 'interactive novelty seekers' were significantly more satisfied with their 'local experience' than 'home seekers'. Satisfaction with the city did not significantly differ between Airbnb users and hotel guests, with both groups mentioning similar elements or experiences as the reason for their satisfaction (Van den Broek, 2017).

Summary: Price, Space and Amenities

Despite the lack of conclusive evidence because of the unavoidable use of convenience sampling in all these studies, a number of observations can be made as directions for further research. In the first place, Airbnb users seem to be mainly leisure travellers, who hardly show any differences in their tourist behaviour when compared to those who stay in more traditional accommodation. Although early research sought to identify the characteristics of Airbnb guests as a practically homogeneous group, Guttentag's study has demonstrated that there are clearly distinguishable motivational segments. Price is consistently mentioned as a reason to choose Airbnb rather than a hotel. Tangible advantages are often given as a reason in second place, an aspect that is often overlooked, maybe because it does not correspond strongly to the marketed image of the platform. Airbnb is a popular option for families on city trips; this is a relatively recent phenomenon for which hotels, probably, have not come up with a suitable alternative. Contact with locals, off the beaten track travel and experiencing new things are other reasons, but apparently of less importance than price, space and amenities.

Note

(1) See https://twitter.com/airbnb/status/694189087238959106.

5 Adventurous Start-up or Real-estate Tycoon: The Airbnb Host

> Airbnb helps Barcelona families earn extra money to stay afloat during these difficult times by opening their homes to visitors.
> General Manager for Airbnb Spain and Portugal Jeroen Merchiers
> (Airbnb, 2013e)

> Every study of Airbnb data that's not commissioned by Airbnb comes to the same conclusion that the image of 'everyday people trying to make ends meet' is nothing but window dressing. It's not a matter of big, illegal hotels on the site, it's a matter of slick, professional hosts who see an opportunity to build a mini empire.
> Jason Clampet, Skift (Clark, 2016)

As argued in Chapter 2, Airbnb caters to two different groups of clients. The assumption about a so-called peer-to-peer (P2P) network is that both parties involved in the transactions are 'peers', and therefore have similar characteristics. In this chapter we will show that this is not necessarily the case, even though it is even harder to collect host profile data than it is to obtain information about guests: not only may they be hard to find, but it is not unlikely that they deliberately avoid providing information about their business because of rental regulations or taxation.

When we examined the promotional videos of the platform, we did indeed see hosts and guests of very similar age groups, perhaps with the exception of the Mexican mom drawn into the business by her son. Their motivation seemed based on the pride of the 'unknown', and allegedly less 'touristy', side of their city, country or subculture. The videos show that it is fulfilling to create moments of bliss, allowing visitors to step out of their tourist role. In other campaigns, however, Airbnb underscores that being a host is, most of all, rewarding financially:

> The study found 84 percent of Airbnb hosts in Portland share only the home they live in – their primary residence – and use the money they earn to help afford the increasing costs of living. Sixty-five percent of Portland hosts said they have used Airbnb income to help afford staying in their

homes. Hosting also helps Portland residents to pursue innovative careers and non-traditional forms of work. The study found 45 percent of hosts are self-employed, freelancers or part-time workers, including 12 percent of hosts who have used Airbnb income to support themselves while launching a new business. (Airbnb, 2014a)

Publications such as these are meant to convince city authorities about the beneficial effects of the urban rental phenomenon. It is important to remember that platform growth not only requires attracting more travellers, but also needs marketing efforts on the supply side. In campaigns meant to enlarge their host client basis, the message is even more directly focused on the financial gains, but also in strict relation to self-actualisation, or what we have called 'existential authenticity' in Chapter 3:

> Airbnb has literally allowed me to follow my passion. Designing furniture was just a hobby, I was always a construction carpenter. Two friends of mine were doing Airbnb, both had really positive stories. I decorated the apartment and pulled it up as a private room, and it's been rolling on since then. Now I have extra money to buy material, more time to focus on designing, keep doing what I'm doing while meeting guests and meeting new people. Airbnb has always given me the independence to be a designer, free me from the shackles of construction. (Airbnb, 2015b)

> I was looking for some sort of challenge in life, and I thought, why not climb Kilimanjaro? And with that, all the expenses start piling up. And I thought, well, I have the extra room in the house, why not rent it on Airbnb? The first weekend I made enough money to be able to purchase the boots, the next week the bag … I started hosting so that I could afford to climb Mount Kilimanjaro, I've continued to host to start saving for the next challenge, which will be another peak somewhere in the world. (Airbnb, 2016c)

Popular areas in Amsterdam have seen active campaigns to convince residents to join the platform with tram stop posters and door-to-door flyering (Figure 5.1).

Hosts' Profile

A qualitative study of the monetary aspects of network hospitality, based on 11 interviews with Finnish Airbnb hosts (Ikkala & Lampinen, 2015) again manifests this combination of motives. The demographics as described correspond to the 'millennial' image of hosts; moreover, they display selective behaviour with a preference for guests who are like themselves. The money is an important driver, as an extra but also as part of a host's monthly budget to pay the rent. At the same time, the hosts indicate that they enjoy the sociable interaction. This pleasure may be similar to that of professionals who work in hospitality, especially in the case of 'remote hosts' who sleep somewhere else when they have guests: in these cases, hosts perform 'small acts of hospitality' (Ikkala & Lampinen, 2015: 1038) such as preparing some food or giving advice, and find their

Figure 5.1 My spare room helps me start my own business. My house paid for my kite board. Earn extra money by sharing your home in Amsterdam.

Source: http://www.floc.nl/productions/airbnb. Every effort has been made to contact the copyright holders and obtain permission to reproduce this material.

gratification in an equally remote response: a letter or a kind review. Host–guest interaction remains limited to the time of stay, and does not lead to longer lasting friendships.

Research in the Netherlands has shown that positive attitudes towards sharing and Airbnb outnumber negative reactions. This preference is skewed for respondents with higher education, high income and progressive or liberal political points of view. Despite the positive attitudes, only 1% participate as hosts in home sharing, and 2% as guests (Beer & Gier, 2016). Further insight into hosts' demographics and motives is anecdotal and mainly comes from non-academic literature (Corbett, 2015; Dobbins, 2017; Garrett-Price, 2014; Peltier, 2015).

Multilisters

With the image of the Airbnb host community apparently determined by the marketed image of authenticity-seeking cosmopolites and independent spirits, the emergence of greedy landlords and illegal pensions on the platform was a shocking discovery. Activist researchers such as Murray Cox (and his website InsideAirbnb) and Tom Slee were among the first to bring this phenomenon to general public attention (Cox, 2015; Slee, 2015). Airbnb has responded to these accusations by claiming that close to 90% of the hosts on the platform offer their own residential property (Airbnb, 2012, 2013a, 2013b, 2013e). This is a somewhat deceitful statistic because, obviously, if 10% of hosts offer more than one property they represent much more than a 10% share of the listings offered. In other words, Airbnb conceals an unequal situation in which a few hosts offer many properties, and many hosts represent a relatively limited part of the offer.

Numerous international studies have confirmed the important share of hosts with multiple listings (so-called 'multilisters') on the platform, either by using information obtained under subpoena from the platform itself (New York State Attorney General, 2014) or with scraped data enabling the connection of properties to single host IDs (Bakker *et al.*, 2016a, 2017a, 2018b; O'Neill & Ouyang, 2016). The phenomenon of multilisters implies that in many cases Airbnb is no longer about sharing one's home to make a living, but about exploiting urban housing to make a profit.

Many of these multilisters appear to buy or rent residential properties for the purpose of turning them into vacation rentals. A look at host profiles suggests that the typical multilister of this type is a financially successful individual who invests his earnings – or previous Airbnb revenues – in residential real estate rather than in stocks or a savings account. Slee gives the example of the owner of a picturesque property in Rome's Trastevere:

> He does have family roots in Rome, but is a Harvard-educated technology entrepreneur who lives in Austin, Texas and who is 'rent[ing] out the places I bought with the proceeds from my last software company.' He is now CEO of Vreasy, a 'novel software platform' which is 'a growing force in the tourism and travel market.' Martin is listed as an Airbnb 'superhost.' In addition to the Historic Nobleman's Loft his listing page shows (in April 2015) a total of six listings on the site – one more in Rome (an 'Ultimate Panoramic Roman Penthouse'), one near Monaco, two in Barcelona, and one 'Seaplane cabin' that he will fly to any legal European lake so that you can use it as a hotel. (Slee, 2015: 49)

Our analyses consistently show that, in Europe's capitals, the share of listings offered by multilisters varies from one-quarter to about two-thirds of the entire offer (see Chapter 1, Table 1.3). Local differences are explained by a number of factors such as the return on investment – average rates are considerably higher in expensive hotel cities – or the speculative character of the housing market. In cities such as Madrid, Zagreb or Tallinn, the share of properties offered by multilisters was found to be 63.1%, 52.4% and 50.0%,[1] respectively. The same data were an indication that multilisters outperform hosts with a single listing.

The evolution of the multilister shares suggests a growth dynamics of successful hosts expanding their business by incorporating further listings, as shown for London in Table 5.1. The accumulative process becomes even more clearly visible if we zoom in on the 3060 London Airbnb hosts who operated a single listing in 2012. Tables 5.2 and 5.3 show how these hosts gradually incorporate additional listings over the years, with 29 out of the original 3060 holding 679 listings in 2017.

Table 5.1 Evolution of host share per category in London (2012–2017)

	2012		2013		2014		2015		2016		2017	
1 listing	3060	60.4%	5907	59.1%	13,365	57.7%	33,016	53.9%	56,226	48.3%	77,463	42.9%
2 listings	882	17.4%	1736	17.4%	3834	16.6%	10,150	16.6%	19,472	16.7%	29,058	16.1%
3–10 listings	848	16.7%	1673	16.7%	3905	16.9%	10,632	17.4%	22,112	19.0%	37,698	20.9%
More than 10 listings	276	5.4%	682	6.8%	2044	8.8%	7413	12.1%	18,639	16.0%	36,154	20.0%
Total	5066	100.0%	9998	100.0%	23,148	100.0%	61,211	100.0%	116,449	100.0%	180,373	100.0%

Source: AirDNA, Bakker et al. (2016b, 2018b).

Table 5.2 Accumulation of properties by 3060 London single-listers since 2012

	2013		2014		2015		2016		2017	
1 listing	2831	84.3%	2668	72.6%	2492	60.2%	2316	49.8%	2168	41.0%
2 listings	348	10.4%	548	14.9%	730	17.6%	932	20.1%	1036	19.6%
3–10 listings	180	5.4%	440	12.0%	790	19.1%	1058	22.8%	1411	26.7%
More than 10 listings	0	0.0%	20	0.5%	127	3.1%	341	7.3%	679	12.8%
Total	3359	100.0%	3676	100.0%	4139	100.0%	4647	100.0%	5294	100.0%

Source: AirDNA, Bakker et al. (2016b, 2018b).

Table 5.3 Number of listings held in 2017 by those who had a single listing in London in 2012

Hosts	Listings	Hosts	Listings	Hosts	Listings
2168	1	7	10	1	23
518	2	7	11	2	25
179	3	3	12	1	46
64	4	3	13	1	47
38	5	2	14	1	55
19	6	1	15	1	73
13	7	3	18	1	95
9	8	1	19		
9	9	1	22	Total hosts: 3053	

Source: AirDNA, Bakker et al. (2016b, 2018b).

To put it cautiously, being an Airbnb host seems addictive. The rapid expansion can be explained by a high return on investment, especially if we look at the highest yielding neighbourhoods in each city (Table 5.4). Of course, revenues per listing vary enormously depending on listing properties and location (London: $SD = 9856$). While it is not uncommon for less attractive listings to achieve not a single sale in a year, the maximum yearly revenue for properties in the popular areas of London or Berlin amounts to €280,000.

Perhaps an important limitation of these multilister analyses, as well as previous studies by others, is that the data are usually collected at destination level. In other words, investors with three houses in London can be identified in these databases, but not those with one property in London, one in Bournemouth and one in Paris. For the purpose of managing an Airbnb operation, working in one city might be the most sensible option; there were no multilisters who combined properties in more than one of the four cities – Amsterdam, Berlin, London and Madrid – we examined in 2016. But some of the Airbnb management businesses are

Table 5.4 Average yearly revenue per Airbnb listing in Amsterdam, Berlin, London, Madrid, Paris (2017)

City	Av. yearly revenue	Neighbourhood	Av. yearly revenue
Amsterdam	€10,602	Centrum-West	€14,635
Berlin	€4056	Mitte	€5075
London	€7590	Kensington and Chelsea	€12,774
Madrid	€6225	Centro	€8104
Paris	€7512	Louvre	€15,059

Source: AirDNA, Bakker *et al.* (2018b).

already internationalising, and Slee's example of the Austin/Trastevere Superhost Martin shows that the platform is also suitable for globalised cherry pickers.

In the literature, the circumstance of a host offering multiple properties – either minimum two (Li *et al.*, 2015) or three (New York State Attorney General, 2014) – is considered as an indication of professionalisation; single-listers are held to be 'inexperienced individuals who list their spare rooms or apartments/houses for rent' (Li *et al.*, 2015: 7). The financial outperformance by professional hosts that we have already observed in the difference between total listing share and revenue share of the different host categories was proven in this study performed in Chicago in 2012–2013, which shows that multilisting hosts earn 16.9% more in daily revenue, have 15.5% higher occupancy rates, and are 13.6% less likely to exit the market. The explanation is sought in the pricing inefficacy of the non-professionals, who do not adopt dynamic pricing strategies to adjust rates to peak events or seasonality (Li *et al.*, 2015). A study on Airbnb pricing in Amsterdam confirmed that frequent price adjustments lead to a better performance (Oskam *et al.*, 2018). On the other hand, Gibbs *et al.* (2018) find lower prices for multi-hosts, an effect they suggest may be due to the fact that commercial providers seek high occupancies, while casual hosts weigh the risk of offering their property.

Professionalisation

Even if the emergence of professional and full-time hosts seems contrary to the utopian and 'sharing' image the platform seeks to promote, it is at the same time an indispensable ingredient in the company's business strategy for growth and enlarging its market share in the urban accommodation market. It is therefore in its interest to push its hosts – not unlike a company seeking to motivate its employees – to high productivity and service consistency. Not only the financial gains make Airbnb addictive; gamification techniques, as has been shown in the case of Uber (Scheiber, 2017), help spur the productivity of platform members.

Airbnb rewards its high performers with an elite status which may produce further financial benefits: the platform indicates that 400,000 of its hosts go 'above and beyond for every guest' and signals this excellence with a 'Superhost' badge. Current requirements for Superhosts are an average rating higher than 4.8, a 90% response rate to inquiries within 24 hours, an activity of more than 10 stays per year and no cancellations. In return, Superhosts receive better exposure (as searches can be filtered for this category), early access to new programmes and initiatives, and they become eligible for the Airbnb Plus programme of high quality and comfort offerings. As a result, Superhosts can expect 22% higher revenues according to the platform's site (Airbnb, 2018a). Gunter (2018) finds that commercial providers, i.e. multilisters, are more likely to obtain Superhost status, which he explains by their professionalisation – the probability of 'having an Airbnb account in good standing' – and by scale advantages: they are more likely to achieve the required number of bookings.

In their study of price effects of trust indicators, along with other characteristics of the offer, Wang and Nicolau (2017) find that Superhost status raises prices by 8.73%. Xie and Mao (2017) find that host attributes such as Superhost status, operating experience and response rate are positively associated with number of reservations, an association that is mitigated if the host operates multiple listings. A positive effect of the Superhost badge on the number of reviews, ratings and guests' willingness to pay was also shown by Liang *et al.* (2017) for Hong Kong in 2015. Nevertheless, these authors argue that obtaining and preserving the badge requires a considerable time investment for the host in view of the inquiry volume. Further rewards to excellent hosts can be expected to maintain the platform's incentive to professionalisation, such as an invitation to real-estate owners to list their spaces in order for these to be managed by experienced Superhosts (All about Airbnb, 2016).

Concierge Companies

An important category of professional hosts consists of commission-based management companies that handle the marketing, sales and operational aspects of Airbnb rentals for individual residents, multilisting investors or perhaps even corporate owners – for the identity of the owners of managed properties remains concealed in the databases. These concierge companies are included in the databases as 'multilisters', with sometimes several hundreds of listings. Their business model is, however, different from that of the investor-host. Even though it is impossible to distinguish these categories in the statistics, this is the reason why we use different categories or tiers of multilisters – three to 10 or more than 10 listings – to give an impression of the presence of different types of operations.

As with other players in the urban vacation rental market, information about concierge services is not easily available. By mid-2016 the

Netherlands had around 22 of these start-ups, concentrated in Amsterdam. They varied in size from one or two people taking care of a few houses to 20 employees managing over 600 listings. With brand names such as Iambnb, Airbnbutler, Bnbmanager, KeyOkay, The Friendly Host, Holiday Host, 60 Days or AwayKey, the businesses seemed in general to be owned by recently graduated millennials with various business degrees, expressing a strong belief in the 'sharing' economy and, following our calculations, earning an above-average salary.

The general standard, despite the highly competitive environment, is an all-in fee of 25% of the rental earnings charged to the home owner or resident, with no additional cost for the guest. By optimising rental prices and occupancy – most concierge services use competitor platforms alongside Airbnb – they claim revenue increases of up to 40%. Although generally associated with the excesses of Airbnb rentals, representatives of concierge services surprisingly welcome stricter legislation, as it avoids a further professionalisation that would make their services superfluous. Their success in Amsterdam has made some of these companies explore international expansion of their businesses (Harboe Sorensen, 2016).

Multilisters, Superhosts and concierge services are signs of a professionalisation that can potentially change the urban vacation rental market from heterogeneous small-scale suppliers to large commercial investors repurposing residential real estate. Despite the platform's strategy to get developers and facility-management companies on board, this type of multilister has been reluctant to enter the market, according to a professional investment blog because of uncertainties about municipal regulations:

> Companies like AvalonBay and Camden Property Trust own tens or hundreds of thousands of units, and they spend hundreds of millions of dollars buying and constructing residential buildings. These companies normally rent out apartments to people who sign year-long leases. But they could instead rent them out on sites like Airbnb. (Realty Shares, 2016)

A further move in this direction was seen in 2017 with the announcement of a 324-unit apartment building in Kissimmee, FL, in cooperation with a local developer (Newgard Development Group), to be operated as a kind of timeshare under the brand of 'Niido powered by Airbnb' (Carson, 2017; Zaleski, 2017a).

Working Relations in the 'Sharing Economy'

The so-called 'sharing' or 'gig' economy has profoundly altered labour relations through its matchmaking capacity between supply and demand: for some businesses, productivity no longer depends on sustained employment relations, but can instead rely on ephemeral contracts with a global workforce. According to those who have promoted 'sharing', this creates a micro-entrepreneurs' paradise in which innovative ideas flourish once people are freed from oppressive nine-to-five schedules (Sundararajan, 2016); for others, it erodes the negotiating power and conquered rights of the workforce and turns its members into 'permanent hustlers' (Morozov, 2014). This debate has mainly revolved around platforms where people offer their labour power, such as task platforms – Amazon Mechanical Turk, Taskrabbit, ListMinut.be – or Uber.

Advocates of the first vision will argue that workers' dependence on alienating jobs keeps them from investing their time and creativity in pursuing their real passion. Sundararajan quotes a copy shop worker who finally decides to quit her full-time employment to start a profitable hand-crocheted scarf business on Etsy:

> There was a point when the amount of hours spent at my place of employment began to actually reduce the amount of total income being earned through my 'secret projects' on a then relatively unknown site called Etsy.com. (Sundararajan, 2016: 106)

The feared race to the bottom does not occur, however, when talented craftsmen leave an unskilled job to pursue their vocation, but rather when regular employment is replaced by precarious freelance jobs through internet platforms. Not unlike day labourers waiting at the village square during the harvest season, workers of different skill levels flock together on platforms from Mechanical Turk to TakeLessons with their newly conquered freedom to sell their labour force. A crocheted scarf business with over US$140,000 in yearly revenues is hardly representative of this new 'precariat': a Belgian study (De Groen *et al.*, 2016) showed that 95% of those looking for work on a task platform did not earn a single Euro. Of those who successfully found a job or task, most made between €1 and €100. The maximum amount earned by a single worker was €5663. The study therefore concludes that 'the platform does not provide sufficient income to be a credible substitute for an offline job' (De Groen *et al.*, 2016: 19).

Nevertheless, an International Labour Office (ILO) survey (Berg, 2016) found that 40% of the workers on task platforms relied on these platforms as their main source of income. Globalisation drives wages down, as workers from high-income countries compete directly with those from cheap labour countries on the platforms. Precarious workers lack the bargaining power of labour representations, and may be fired – 'deactivated' – from platforms such as Uber because of underperformance, but also allegedly for disciplinary reasons because of a critical attitude towards the company (Huet, 2014).

It is often suggested that labour flexibilisation satisfies the aspirations and career visions of a millennial generation; such suggestions are of disputable evidence and raise a chicken-and-egg question. Wage stagnation and decreased job security are employer-driven developments (Meek, 2014), to which contemporary workers must respond with a permanent alertness to job opportunities. Simply put, millennials' decreased job loyalty corresponds to companies' decreased worker loyalty. It is therefore hardly surprising – although counter-factual – that for a majority of millennial workers the main appeal of the 'gig economy' is the desire for a higher income than in their current jobs (Deloitte, 2018).

Entrance Barriers

Access to the means of production is a further complication in the gig economy. Unlike the factory worker who has nothing to sell but his or her time and labour force, the gig worker is expected to bring his or her own tools and machines. This entrance barrier refutes the emancipatory illusion that the sharing economy would somehow subvert social inequality, as argued by Sundararajan:

> In a sense, with little additional infrastructure, Piketty's 'renters' can begin to experience the other side of the coin by making money through investing or owning rather than laboring. (Sundararajan, 2016: 124)

The more one can invest in computational power, the more bitcoins can be mined; to become a Lyft driver, one's car must comply with the company's standards (Dillahunt & Malone, 2015). And while Airbnb does not vet its hosts or their properties, the demand for units in less privileged and peripheral neighbourhoods can be expected to be inferior to the appeal of luxury apartments in tourist areas, as will be analysed below.

This inequality is aggravated by the circumstance that some micro-entrepreneurs participate in the gig economy to pay their debts or to make ends meet, while for others their business means additional income. This may seem a trivial difference in the initial phase of the platforms, but it is likely to determine the entrepreneurial careers of gig workers. Sundararajan celebrates another success story:

> In 2013, a Turo member named David learned about the platform and decided to rent out a truck he was about to sell. The experience was positive enough for David to start using Craigslist to purchase other vehicles to list for rent on Turo. As of 2014, David had six vehicles and, sharing his story with a Turo blogger, he reported earning a couple thousand dollars per month from his microfleet. (Sundararajan, 2016: 107)

In other words, in the gig economy some micro-entrepreneurs also get a head start – let's call this 'primitive accumulation' – that allows them to build the 'mini empires' to which Clampet's opening quote of this chapter referred. In our analysis of the multilister phenomenon in London (Tables 5.2 and 5.3), we have seen how a limited number of Airbnb hosts, similarly to David with his trucks, have succeeded in accumulating an increasing share of Airbnb units.

The Work of a Host: Setup, 'Overheads' and Booking-related Tasks

The situation of urban vacation rental hosts is different from that of Uber drivers or platform workers, in the sense that their revenues are generated by their available capital – houses and rooms – rather than their labour. Nevertheless, their income has a labour component, which we can break down as follows: initial setup efforts (preparing the apartment, the platform account and website details); a permanent sales activity (setting prices, making reservations, responding to clients); and variable activities per booking (receiving guests and cleaning the apartment). Although there are no specific data to quantify these activities, the ILO study found that platform workers average 28.4 working hours per week, of which they dedicate 6.6 hours – or 23.3% – to 'overhead' time in order to acquire and manage jobs or, in other words, 'for every hour of paid work, workers spent 18 minutes searching and doing unpaid preparatory work' (Berg, 2016: 11). It seems plausible to assume that urban vacation rental hosts also dedicate a substantial amount of time to this type of 'overhead'.

The variable workload per booking equals the operational tasks in traditional hospitality: mainly welcoming the guest, checking in and housekeeping, usually limited to a single cleaning after the stay. Concierge companies will professionalise this service, but for smaller operators the work volume is probably too limited: the average London single-lister on Airbnb had six bookings in 2017, increasing to 12 bookings per unit for a multilister (Bakker *et al.*, 2018b). While the single-lister may clean the house him- or herself, an investor with 10 listings will have 120 cleaning jobs per year, which can be outsourced, possibly in the informal sector. As a consequence, we may say that the micro-entrepreneurial activity generated by the 'sharing' platforms comes at the expense of regulated labour. A study on the impact of Airbnb in Los Angeles makes the following estimate:

> If AirBnB units were hotel rooms, the 11,401 units on the Los Angeles market would employ more than 7400 hotel workers, earning an average wage of $14.07 per hour. However, one way AirBnB keeps overhead low is to outsource traditional hospitality labor jobs, most notably housekeeping. Housekeeping is likely carried out by domestic workers employed by any number of home cleaning services. Domestic workers earn a median wage of $10 per hour. (Samaan, 2015: 15)

The same study further estimates the total job loss, due to less frequent cleaning and the suppression of other hospitality jobs such as front desk staff, valet parkers, telephone operators, shuttle drivers and security and janitorial staff, to amount to 80% of the equivalent 7400 positions. On top of that, the competition of urban vacation rentals may undermine the job security of the remaining workers in the traditional hotel sector.

But the micro-entrepreneurial climate is still in the earliest stage of its evolution. We have already alluded to signs of business concentration on 'sharing' platforms. The constantly required 'overhead' efforts, in particular, mean that hosts with multiple units work more efficiently than single-listers, a scale advantage that is reinforced by a learning curve through which a host becomes better prepared to select guests, set prices and perform other managerial tasks (Dolnicar, 2018; Oskam et al., 2018). We can therefore suppose – experimenting with different hypothetical time quantities for the three work components of hosts – that there is a linear relation between the number of units operated by a host and revenues per hour worked. Put differently, multilisters' operations become progressively more efficient. As the market matures, we may therefore expect scale increases, because single-listers abandon the business, because they outsource their activities to concierge companies, or because they liquidate the increased value of their property by selling it to other operators.

In short, the micro-entrepreneurial utopia envisioned by advocates of the 'sharing' movement may well be a transitory stage after which ordinary people selling their labour and their assets make way for ever bigger concentrations of capital. These accumulation and professionalisation processes will not just be decisive for the future of the platforms; they will also lead to a shift from regulated employment to precarious jobs, and to a dispossession of the means of production – urban residential housing.

The Empowerment of Poor Residents

The platform's claims of supporting the ordinary man in making a living are undermined by the aforementioned barriers to entrance, and the subsequent processes of accumulation and professionalisation. Simply put, those who are able to afford bigger houses in more popular neighbourhoods will also be able to attract more guests and at higher rates.

Studies of Airbnb's spatial distribution in cities consistently confirm this assumption: most listings are concentrated in central and touristic areas (Arias-Sans, 2015; Bakker et al., 2016a, 2017a; Quattrone et al., 2016; Slee, 2015), with distance to the city centre negatively correlated to Airbnb prices for all studied cities, with the single exception of Budapest (Dudás et al., 2017). An international study finds a 0.59% reduction per kilometre, in general, with a stronger effect for the higher end listings (Wang & Nicolau, 2017): 3.3% in Canadian cities (Gibbs et al., 2018); 5.11% in Rotterdam (Koomans, 2018); or US$5.29–7.63 per kilometre in Germany (Teubner et al., 2017). In general, we can also observe higher rates for bigger properties, as shown in Table 5.5.

An interesting insight into how Airbnb's growth dynamics strengthen socio-economic inequality is provided by a longitudinal study of the platform's spreading patterns in London from 2012 to 2015 (Quattrone et al., 2016). The study shows that in 2012 the main predictor for the appearance

Table 5.5 Average daily rate for Airbnb listings of different sizes in Amsterdam, Berlin, London, Madrid, Paris (2017)

2017	Shared room	Private Room	Entire home, 1 bedroom	Entire home, 2 bedrooms	Entire home, 3 bedrooms	Entire home, 4 or more bedrooms
Amsterdam	€94	€97	€129	€184	€244	€331
Berlin	€28	€38	€66	€99	€152	€233
London	€30	€56	€115	€174	€248	€326
Madrid	€25	€32	€70	€102	€140	€241
Paris	€32	€56	€80	€153	€245	€385

Source: AirDNA, Bakker *et al.* (2018b).

of Airbnb listings was geography – closeness to the city centre – and that these first hosts were probably young and ethnically diverse residents of central neighbourhoods, with a strong presence of students suggested by a negative correlation with employment. In 2013 Airbnb spread to areas with home owners and lower income residents. Finally, in 2014 and 2015 low-income and rental homes emerge as the strongest predictors, suggesting that participation is driven by financial motives. However, the demand growth does not reflect this distribution pattern as it remains concentrated in central areas, leading to the conclusion that later adopters driven by financial needs do not benefit from the growth of the platform: 'many properties that are listed, but are too far from touristic areas, are not being rented out' (Quattrone *et al.*, 2016: 7).

Finally, painful evidence against the empowering claims of the platform are differences in revenue associated with the racial background of hosts, differences that may be explained by discrimination or by the associated socio-economic differences. Black hosts in New York have been found to receive significantly lower average prices than non-black hosts (US$107 versus US$144; Edelman & Luca, 2014), while Hispanic and Asian hosts have lower list prices than white hosts in San Francisco (9.6% and 9.3%, respectively; Kakar *et al.*, 2017), or up to 20% for Asian hosts in Oakland-Berkeley (Gilheany *et al.*, 2015). A 2017 study in New York found Airbnb listings in predominantly black neighbourhoods to be mostly operated by white hosts, making urban vacation rentals an important factor in the displacement of the original population (Cox, 2017a).

> **Summary: Free Market Equality**
>
> Anyone can become an Airbnb host. This generous invitation is what the empowering claim of Airbnb boils down to. But once on the platform, unrestricted market forces favour some more than others.

> People with large properties, people with properties in central and attractive locations, owners with many properties and experienced professionals with market insight will outperform the ordinary man in the suburbs. This advantage will allow them to accumulate capital and to concentrate market power. Urban vacation rentals thus contribute to social inequality and have even been surprised by illegal discriminatory practices.

Note

(1) Tallinn: data for 2016.

6 The Established Order: Impact on the Hotel Market

> I was not surprised that you are unwilling and unable to defend your industry's longstanding commitment to price gouging consumers, depressing wages and replacing workers with robots. Last week, we released a new report highlighting your industry's habit of taking billions of dollars from taxpayers to subsidize the construction and operation of your hotels. The report finds that Marriott received $1.73 billion in taxpayer subsidies between 2008 and 2017. From 2008 to 2016, you have generated $4.6 billion in profits.
>
> Letter by Josh Meltzer, Head of Public Policy at Airbnb, to Arne Sorenson, CEO Marriott (Bosa, 2017)

Just as Airbnb has convinced travellers that they are not tourists, the world's largest accommodation distributor has not only successfully conveyed the message that it is not part of the hotel industry, but it has framed its competitive position as a Manichean fight against the powers that be. The 'sharing' narrative and the 'culting' practice have succeeded in outmanoeuvring the traditional marketing techniques of the opponents targeted at their believers, which 'will easily be interpreted as attempts to deter them from the "path of righteousness"' (Strong, 2014).

It has certainly been helpful that the traditional hotel companies have waged a war on the disruptive intruder by calling for a 'level playing field', or equal regulatory conditions regarding taxation, safety, labour laws, etc. Their strategy aimed to 'ensure comprehensive legislation in key markets around the country and create a receptive environment to launch a wave of strong bills at the state level while advancing a national narrative that furthers the focus on reining in commercial operators and the need for commonsense regulations on short-term rentals', according to documents of the American Hotel and Lodging Association (Benner, 2017a).

Market Share

Airbnb claims to generate at least partially incremental demand for city visits: 61% of Airbnb clients visited Barcelona for the first time (Airbnb, 2013e); 27% would not have come to Paris or stayed as long if it were not for Airbnb (Airbnb, 2013a); and 51,000 visitors to New York (or 7% of Airbnb users) would not have stayed in the city without Airbnb (Airbnb, 2015a). It is hard to verify these claims as the growth of Airbnb can be seen as part of a wider trend of spectacular growth in city trips: 82% between 2007 and 2014, to reach a 22% share of all holidays. Therefore, city tourism has constituted the dynamic segment which has been the great driver behind the general growth of the sector, according to IPK (2015). Evaluating whether the Airbnb share has been incremental or at the expense of traditional hotels is even harder if we consider that international visitor numbers are usually based on hotel statistics. Morgan Stanley has reported that 42% of Airbnb users substitute away from hotels (Nowak et al., 2015). The relative market shares of Airbnb can be calculated by comparing overnight stays in Airbnb accommodation with hotel nights as in Table 6.1.

Table 6.1 Market share of Airbnb in overnight stays: Amsterdam, Berlin, London, Madrid, Paris, 2016–2017

	2016		2017	
	Airbnb Booked nights	Market share	Airbnb Booked nights	Market share
Amsterdam	1,662,000	10.7%	2,100,000	11.8%
Berlin	1,735,000	5.3%	2,160,000	6.5%
London	4,619,000	5.0%	6,703,000	6.9%
Madrid	1,290,000	6.7%	2,155,000	10.1%
Paris	5,020,000	13.4%	6,449,000	15.2%

Source: AirDNA, Bakker et al. (2018b).

Real-estate advisor Colliers expects the Airbnb market share to eventually stabilise at 12–14% of overnight stays. The overall growth of visitor numbers to cities means, in most places, that hotels apparently do not suffer from Airbnb: tourism is booming and hotel performance has improved in recent years, as indicated in Table 6.2 by revenues per available room (RevPAR).

This also means that a loss of market share may become visible once volumes stop growing, because of an economic downturn or other specific circumstances. Airbnb demand growth is such that it not only compensates for seasonality effects, but it also contrasts with a temporary setback in the London hotel market in 2016 due to supply growth and exchange rate fluctuations, as shown in Table 6.3. Airbnb is affected by deeper crises such as the one caused by the terrorist attacks to Brussels, but whereas this

Table 6.2 RevPAR evolution in European cities

	2010	2011	2012	2013	2014	2015	2016	2017
Amsterdam	89	89	88	90	92	102	105	116
Berlin	59	59	63	63	66	71	72	73
London	127	125	128	128	132	133	130	137
Madrid	56	58	55	51	49	48	69	73
Paris	161	184	199	201	197	184	157	170

Notes: Hotel RevPAR (revenues per available room) calculations differ considerably per source; STR findings are on average 4.5% higher. The data in Table 6.2 are meant to indicate general trends, rather than the most accurate numbers per year.
Source: PwC, Statista.

Table 6.3 Evolution of hotel and Airbnb performance, London

	Hotels			Airbnb		
	January 2015	January 2016	Year-on-year change	January 2015	January 2016	Year-on-year change
Demand	2,640,900	2,590,400	−2%	74,900	229,100	206%
Revenues	€414,672,245	€378,756,940	−9%	€8,604,125	€24,290,025	182%
Occupancy[a]	70%	67%	−5%	9%	21%	126%
ADR	€157	€146	−7%	€115	€106	−8%
RevPAR	€110	€98	−11%	€11	€22	108%

Notes: [a]Occupancy, in the case of Airbnb, stands for reserved days/(reserved days + available days). This indicator is hardly comparable to hotel occupancy, as most listings cannot be rented out permanently. The same goes for RevPAR (revenues per available room).
Source: Bakker et al. (2016b).

led to a decline for hotels, the consequence for Airbnb was a slowdown, indicated in Figure 6.1.

Still in the early years of urban rentals, Zervas et al. (2017) quantified the revenue loss on the Texan hotel market, with an estimate of 13% loss of room revenue for Austin and a 0.35% decrease in monthly hotel room revenue for every 10% increase in Airbnb listings for Texas in general. This study observed that these effects were especially noticeable at the lower end of the market, in hotels that lacked business facilities. Similar competitive effects were found in a study on the effect of Uber on taxis in New York and Chicago. The growth of Uber would lead to a reduction in complaints for regular taxis, which is interpreted by the authors as a clean-up of the system: taxis are forced to improve quality or they are driven out of business (Wallsten, 2015).

Despite increasing average daily rates (ADR), especially in expensive hotel cities, the majority of the Airbnb offer should still be considered low cost, as the median ADR lags considerably behind the mean: €89 versus

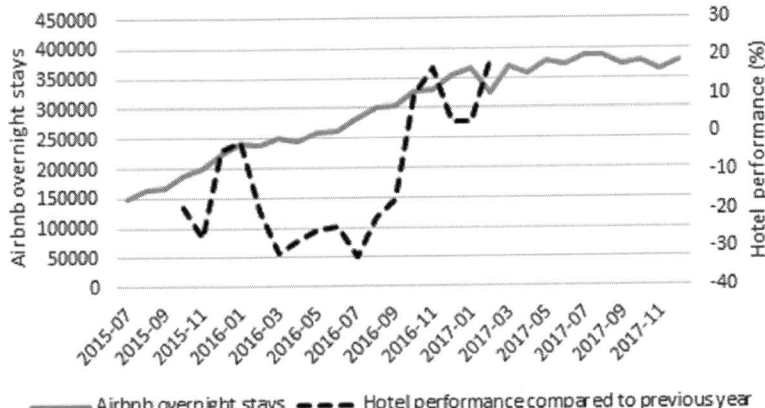

Figure 6.1 Airbnb performance during hotel crisis, Brussels
Source: Bakker *et al.* (2018b); Business Travel IQ (2017).

€110 for London, and €126 versus €135 for Amsterdam in 2017.[1] In Chapter 4 we discussed the attempts of the platform to enter the business market, which we consider not to be entirely successful. Upmarket hotels especially fear competitive effects during peak times: the use of Airbnb during events diminishes demand compression and, therefore, has a downward effect on the premium rates hotels can charge during those times, according to Laura Paugh, Marriott International's SVP of Investor Relations (Ting, 2017a). However, both Morgan Stanley and a study by STR – based on non-disclosed Airbnb data – minimise these effects, showing, in particular for hotels in US cities, a sustained high number of compression nights with an average 35% price premium, despite a certain volatility due to external effects (Haywood *et al.*, 2016; Nowak *et al.*, 2015).

Hotel Companies' Competitive Strategies

Initially, the strategic answer of traditional hotel companies and their representatives – the American Hotel and Lodging Association, the British Hospitality Association and their equivalents around the world – focused on a legal battle against urban rental platforms, demanding an end to unregulated practices and a competitive 'level playing field', meaning that the platforms would be subject to the same taxation, safety requirements, labour laws and other regulations as hotels. Their lobby supported the deconstruction of 'home sharing' as an idealistic and non-commercial activity. It also sought to nuance the claimed beneficial economic effects of urban vacation rentals, for instance by denouncing the replacement of regulated employment by unpaid host labour. Although this public relations offensive has contributed to a better understanding of 'home-sharing' practices, the legalist strategy has not been very successful. Where it succeeded in convincing authorities to impose stricter regulations, difficulties in controlling the phenomenon often made those rules ineffective. More importantly, just as the music industry had experienced a decade earlier, the legal battle was unable to suppress a consumer trend, as long as the traditional providers could not sufficiently incorporate innovations.

Most hotel chains deny any serious impact from urban vacation rentals, as these are supposed to operate in a different customer segment. A number of tactical and operational improvements have been identified for hotels to adopt from the platform, or at least to be considered as inspiration. User-friendly web design highlighting location and prioritising images over text are basic lessons learnt from Airbnb. Intelligent data use can improve insight into travel habits and shifts in consumer preferences, with the 'home field advantage' for hotels of collecting much more data on their guests' spending during their stay. Also, a better understanding of the need for personalised or customised rooms, including a 'choose your room' option for those who check-in online, is becoming more common. A more general trend seems to be the interest in 'local' or 'authentic' experiences (Forgacs & Dimanche, 2016).

A small number of innovative hotel concepts precisely addressed these apparent preferences of the Airbnb customer by incorporating the platform's 'cool', at least if we follow the partial analysis of Airbnb guest motivations, coinciding with the platform's marketed image. The new concepts mainly appealed to Guttentag's (2016) two groups of novelty seekers: the pragmatic novelty seekers, through the design and amenities offered in the hotel; and the interactive novelty seekers, by introducing innovations to foster communication and community feelings among guests and between guests and employees.

Without pretending to give an exhaustive overview of these new concepts, some innovations clearly illustrate this trend. The first Room Mate Hotel was founded in Madrid in 2000 by Kike Sarasola to become an international chain as of 2005. Its concept was meant to offer hospitality like 'staying with a friend'. Personalised service, central location and attention to design contrast with reduced traditional hospitality services such as bar, gym and pool, thus adapting the offer to the needs of the experienced traveller whose purpose is to enjoy the city and its nightlife (Emprendedores, 2009; Katz, 2012).

The Dutch CitizenM brand – for Citizen Mobile – was created simultaneously with Airbnb, in 2008, as another pioneer of combining affordable accommodation with high-end design. The creator, Rattan Chadha, was the former owner of a fashion brand seeking 'great style for a price a 25-year-old designer could afford'. The prefab rooms offer high-quality essentials – bed, shower and wifi – while reducing space and eliminating extras such as minibars. The common spaces, on the contrary, became more welcoming and better equipped than was usual up to that time, addressing the needs of the company-seeking traveller (Welch, 2014).

Described by news media as 'a combination between Airbnb and a WeWork coworking space' (Garfield, 2016), Hans Meyer's Zoku Hotel in

Amsterdam targets short-stay expats. It continues the trend of reducing traditional hotel services, innovative room design and attention to public spaces. These spaces are designed for hotel guests and non-guests to meet and socialise, and for young and creative professionals to work together.

Explicitly 'geared to the tech-savvy Airbnb set', the PUBLIC New York, created by Ian Schrager, boasts a similar promise of 'luxury for all'. Attention to design, common spaces designed to foster social interaction and non-traditional food and beverage outlets are other characteristics PUBLIC has in common with CitizenM and Zoku (Yuan, 2017). The role and preparation of the hotel staff has shifted from traditional operational activities to improving the guests' experience: the so-called 'ambassadors' for CitizenM or 'advisors' for PUBLIC emulate the personal attention of Airbnb hosts. Rather than employing traditionally trained hotel staff prepared to follow standard operating procedures, CitizenM's 'castings' select people for genuine behaviour and guest attention (Welch, 2014), thus reflecting a wider trend in contemporary hospitality (Dekker, 2014, 2018; Lodder, 2002).

These four concepts are examples of initiatives by innovative entrepreneurs seeking to radically change the hotel business. They replaced professional service with personalised service, and traditional amenities with attention to those elements that closer correspond to the needs of the contemporary urban traveller. They revalued the hotel's public spaces, encouraging interaction between guests, which ironically is an option rarely offered by Airbnb. These concepts appeal to a similar kind of authenticity – both in service offered and in existential guest authenticity – to that marketed by Airbnb. But they compete only to a more limited extent on the tangible advantages of urban vacation rentals that attract guests driven by other motivations, described by Guttentag as 'money savers' and 'home seekers'. Gunter and Önder (2018) recommend that hotels specifically increase their offer of rooms sought after by Airbnb travellers – with local 'look and feel', household amenities and bigger rooms with capacity for more than two guests – while emphasising advantages like their professional responsiveness or shorter minimum stays. 'Airbnb guests demand room types in short supply in the traditional accommodation industry, while preferring a hotel-like booking process' (Gunter & Önder, 2018: 20).

'Branded marketplaces' were suggested in a scenario study as a competitive strategy for hotels to offer urban vacation rentals with consistent service, with the example of a hypothetical *Hiltonbnb* (Richard & Cleveland, 2016). Hotel companies entering the home-sharing market with a hotel-like service level could indeed offer an alternative for those motivational segments attracted mainly to the tangible advantages of an apartment over a hotel room. An example of this trend is onefinestay, an upmarket rival for Airbnb founded in 2009. Its labour-intensive business model – because of a higher service level for both hosts and guests, compared to Airbnb – made it impossible to follow the expansion speed of Airbnb, creating a vulnerability for the company: its concentration in London and Paris caused performance to drop after the 2015 terrorist attacks in both cities. This may be an explanation for the fact that the platform was initially acquired by Hyatt in 2014, to be taken over only two years later by Accor. Hyatt continued to explore the home-sharing market by acquiring a stake in Oasis, another luxury platform that offers urban vacation rentals with professional full-time hosts and additional guest services (Ting, 2017b; Whyte, 2016).

The man behind Room Mate, Kike Sarasola, launched BeMate as a home-sharing platform in 2014. This 'apartment rental and hotel hybrid' offers advantages and disadvantages similar to the luxury platforms. On the one hand, it can offer consistent quality through the selection and control of properties on the platform, plus hotel services such as 24/7 customer or concierge services, housekeeping, airport transfers, babysitting, breakfast, luggage storage and late check-out. The downside is that the platform's expansion remains limited to cities with Room Mate hotels, with which BeMate shares these services (BeMate, 2014a, 2014b).

Airbnb Entering the Hotel Business

The development of hybrid models combining urban vacation rentals with full service is a simultaneous trend on both sides. Airbnb has launched several initiatives to penetrate more upmarket guest segments by awarding special badges to properties with certain requirements. First, 'Business Travel Ready' was introduced in 2015, and rebranded two years later as 'Airbnb for Work'. The requirements include wifi, self check-in and a 'laptop-friendly workspace' – whatever that may be – as well as a minimum overall and check-in rate of 4.8, five yearly reviews and an over 90% response rate in 24 hours.

The development of strategies to strengthen consistency in providing hotel-like services is attributed to Chip Conley, founder of Joie de Vivre Hotels. A greater predictability of the offer was identified as a condition to consolidate the platform's position for more affluent and business travellers because, as an Airbnb guest declares, 'When an Airbnb is bad, it's really bad'. The policy change does not only require an upgrade of amenities, but also that 'hosts will act like hotel staff members, meaning they will be courteous and blend into the background'. The standardisation is therefore not welcomed by all hosts as it goes against their understanding of the original 'sharing' philosophy and has also affected their hosting experience: from a 'life-enriching' activity it has become more similar to acting as a hotel employee. Even if adopting the standards is voluntary, a *New York Times* article indicates that 'interviews with more than two dozen hosts showed that many felt pressured to comply' (Benner, 2017b).

In 2016 Airbnb piloted 'Sonoma Select': selected Superhosts in this Californian city were invited to participate in the 'hotel-like' programme, for which the requirements were instant booking, 24-hour check-in, and amenities such as 'local treats, wine, upgraded bath products, and a guidebook'; the website also offered 3D models of the Select listings (Ting, 2016). This was a prelude to 'Airbnb Select', announced in February 2018: again, this is an invitation-only programme, with a long checklist of required characteristics and amenities, although neither instant booking nor the 3D modelling seem to be part of the package. The programme is meant to enlarge the customer base by luring wealthy and older non-Airbnb users to quality-inspected properties on the platform. Besides home inspections, perks for participating hosts are suggestions for home improvements and even options to finance such measures (Ting, 2018a; Zaleski, 2017b).

The next moves in this segmentation strategy have already been announced. Yurts and treehouses will become listed as 'Airbnb Unique'. For holiday villas the 'Vacation Home' badge has been introduced. 'Airbnb Plus' will be the new luxury tier of verified listings that may replace 'Airbnb Select'. Starting with 2000 listings in 13 city destinations, priced at US$250/night and up, Plus targets upmarket hotel guests or, as a travel site puts it bluntly, 'Airbnb Plus is for people who hate Airbnb'. Finally, the acquisition of Luxury Homes by Airbnb in February 2017 meant a further expansion into the higher end vacation segment motivated by 'higher margins and fewer regulatory issues', besides incorporating knowhow about this promising market. Probably the new range of luxury properties will be labelled with its own badge as 'Beyond by Airbnb' (Calder, 2018; Carey, 2018a; Ting, 2017c).

The convergence of urban vacation rentals with traditional accommodation types requires platforms to have a quality care system in place, and whereas competitors such as onefinestay, Oasis or the Chinese Tujia rely on professional property managers in each location, Airbnb's control so far remains restricted to a one-time inspection, followed by host and guest reviews. Options to develop such a system would be to further professionalise the Superhost community or to outsource the on-the-ground services to the concierge companies that have emerged in different destinations. This externalised approach guarantees the scalability of the platform, but it can be questioned whether it will be sufficient for the required consistency of a hotel-like service (Ting, 2018b).

Airbnb as a Distribution Platform

On 13 March 2018 Airbnb published an open letter to boutique hotel and bed & breakfast owners, inviting them to list their rooms on the platform. The advantages offered over other distribution channels – the letter mentions the online travel agents (OTAs) Expedia and Booking – are lower commissions (Airbnb charges hosts between 3% and 5%, compared to 8–25% for other OTAs), no commitment (the platform works without contracts), and the absence of hotel chains in its listings (Airbnb, 2018b). The requirement for professional providers to be listed is that they offer 'unique spaces and personal hospitality' (Airbnb, n.d.), thus excluding '"mass-produced," or bigger, branded hotels', in Brian Chesky's words (Ting, 2018c).

Hotels seem to have had a limited and concealed presence on Airbnb ever since the platform started growing. The use of personal host names rather than company names, and the use of 'private room' listings makes it impossible to systematically detect this type of listing. Based on conversations with hotel managers with Airbnb listings, the general impression is that their presence on the platform is not really meant to increase sales but is rather motivated by curiosity as to how the competitor functions, insight in nearby listings, pricing recommendations, etc. The desire to experiment to attract millennial guests – supposedly the typical Airbnb customer – can also been seen in strange examples such as the upscale Vienna hotel offering beds to share a room at reduced prices shown in Figure 6.2.

Boutique hotels, hostels, guesthouses and bed & breakfasts are currently accepted categories on Airbnb, even though they usually remain listed under a host's personal name. Their share is still minimal. Skift (Schaal et al., 2018) compares the global hotel supply of 16.4 million rooms with the 2.2 million instant bookables on Airbnb – as a minimum requirement to compete with traditional hotels – and 2000 Airbnb Plus listings. Airbnb reports that in 2018, 24,000 hotels are advertising on the platform. The share of hotel-like listings – pertaining to one of the mentioned categories – in European cities, as well as the revenues they generate, is around 1%, with 1175 hotel-like listings for Paris, 2735 for London and 562 for Berlin. Only Amsterdam has a more important presence of 929 hotel-like listings (4.4% of revenue), out of which 868 (or 3.6% of revenue) are B&Bs.

The call to boutique hotels and bed & breakfasts means that Airbnb ceases to be a competitor for (certain) hotels, and instead explicitly enters the competitive field of OTAs. This evolution into a travel intermediary is natural and mirrors the move of Expedia and Booking into the urban vacation rental market, through the acquisition of HomeAway in the former case, or the direct incorporation of apartment rentals on its site in the latter. To what extent will Airbnb be able to disrupt the intermediary market, as it did with city tourism accommodations?

Online travel intermediation itself is a two-sided market characterised by oligopolistic, 'winner-takes-all' competition, with Expedia and Booking as dominant players. This means that the success of a platform in this market depends on its power to attract an exhaustive offer of accommodation providers on the one side, and a large consumer group on the other, in a mutually enforcing dynamics. This mechanism has caused a swift 'reintermediation' in the travel market, after analysts initially expected a 'disintermediation' when internet distribution emerged: hotels were believed, around the year 2000, to be able to skip the 'middle man' in reaching their clients. Intermediaries' market power led to a high revenue share being claimed in sales commissions. These commissions are

Figure 6.2 Sharing a room with other guests in an upscale hotel: A 'dorm' in the Grand Ferdinand, Vienna

sometimes considered as abusive by hotel companies, even though Cornell researchers have suggested that, through the so-called 'billboard effect', these must be seen as an investment in the visibility of a hotel beyond the single sales event (Anderson, 2009, 2011; Oskam & Zandberg, 2016).

For hotels, the direct consequence of Airbnb evolving as a full-fledged intermediary would be decreased market power of the incumbent OTAs. To expand their customer groups, Airbnb and Booking have made opposite choices: Airbnb has kept hosting commissions low (3–5%) by charging a transaction fee to the guest (5–15%), whereas Booking has successfully built its market share by completely subsidising the guest side with commissions charged to hotels (12–20%). This makes it attractive for hotels to advertise themselves on Airbnb, but at the same time guests will preferably make their booking through the subsidised platform. In other words, the commission structure will allow Airbnb to increase its listings rather than its sales of those listings; adjusting commissions will have the opposite effect. A study by Skift Research (Schaal *et al.*, 2018) weighs the advantages and disadvantages of the Airbnb–hotel convergence for traditional accommodation providers as shown in Table 6.4.

Table 6.4 Hotels: Winners or losers? Or both?

Why hotels win	Risks
(1) Increasingly more options in terms of distribution	(1) Additional supply could create pricing pressure
(2) Increasingly more options in terms of marketing and reaching your consumer	(2) Scoop up independents and reduce soft brand demand
(3) Increasing more brand options with which to affiliate	(3) Improved consistency and quality could appeal to business travellers

Source: Schaal *et al.* (2018).

According to the same study, it will be hard for Airbnb to make up for the two-decade head start the OTAs have in building a comprehensive travel offer. The exclusion of big hotel chains – or 'mass produced' hospitality – is an important limitation to the growth of Airbnb's supply side; at the same time, the original hosts may oppose the admission of commercial providers to the platform and leave. Although Booking or Expedia cannot come close to the number of listings on Airbnb, the OTAs are catching up in their inventory of alternative accommodations. Airbnb may benefit from a loyal community, but the OTAs have superior advertising budgets and search engine optimisation knowhow. The competitive advantages and risks are summarised in Table 6.5.

Table 6.5 How Airbnb could win the online travel battle

Airbnb's advantages	Risks
(1) Community	(1) Deteriorating relationship with its hosts; host fatigue
(2) Brand recognition and awareness	(2) Overextending or distracting itself with endeavours that are outside of the core business
(3) More home-sharing inventory than Booking and Expedia	
(4) Better mobile app integration which is not just limited to accommodations	(3) Not enough accommodation supply (both alternative accommodations and traditional hotels)
(5) Better value proposition for hotels who want to use its platform	(4) US-centric executive leadership team

Source: Schaal *et al.* (2018).

> **Summary: Part of the Industry**
>
> At the beginning of this chapter we discussed the image of 'accommodation sharing' as something opposed to the 'hospitality industry'. That distinction is misplaced, as different commercial services in travel accommodation have started to converge. Urban vacation rentals have achieved an increasing share in a city trip market that is also growing. With innovative hotel brands offering more personalised services and paying more attention to human interaction, and with Airbnb professionalising its hosting and seeking greater service consistency, the differences between accommodation types blur. The platforms have also entered into competition with OTAs, but for now their main opportunities seem limited to their niche of alternative accommodation and boutique hotels.

Note

(1) Median ADRs were calculated over the last 12 months since 18 January 2018 for Amsterdam, and 2 February 2018 for London.

7 Suburb Safari or Hotspot Hype? Airbnb in the Neighbourhoods

> We only have to look around us to see how the City and we as its residents are being forced to change. The historic centre is turning into a tourist ghetto from which we, locals, are displaced, grocery stores are disappearing to be occupied by hipster restaurants, ice-cream and souvenir shops, common spaces such as squares are being invaded by terraces, rents and food have seen outrageous price hikes making them affordable only to the wealthy visitor, our streets have become transit routes for the consumer-tourist, and suffer from massive and excessive human pressure that makes life unbearable when the summer begins or when they come down from the cruise-ships ... And still for this year they have forecast a record number of tourist arrivals![1]
>
> Ciutat per a qui l'habita, 2017

Tourism weariness and the emergence of urban vacation rentals have been concurrent phenomena in many cities. The two are associated, although this does not mean that there is a cause–effect relationship. The global growth of tourist arrivals, expected to reach 1.8 billion by 2030 (UNWTO, 2011), is driven by demographics and increased wealth. Besides the economic upturn after 2013, urban tourism was spurred on by the success of low-cost airlines and a changed perception of city life: from dangerous and dirty places in the 1980s they have become appealing destinations to be visited even with the whole family (IPK, 2015). The appearance of low-cost tourist accommodation, through the emergence of urban vacation rental platforms, has accompanied or maybe even catalysed this recent boom.

Spatial distribution is an important element in this discussion. Cities affected by unsustainable tourist pressure or overtourism have identified spreading tourism to non-tourist neighbourhoods or even to other cities as a remedy against the excessive concentration of visitors. This also happens to be a claim that Airbnb advertises in its reports, which would imply that the platform becomes the cure, rather than the disease:

> 72% of Airbnb properties in San Francisco are located outside the central hotel corridor. More than 90% of Airbnb guests visiting San Francisco prefer to stay in neighborhoods that are 'off the beaten track.' Over 60% of Airbnb guest-spending occurs in the neighborhoods in which the guests stay. (Airbnb, 2012)
>
> Airbnb is complementary to the existing tourism industry in Paris. 70% of Airbnb properties in Paris are located outside the central hotel corridor. (Airbnb, 2013a)
>
> 73% of Airbnb properties in Amsterdam are located outside the eight central tourist districts, and 69% of Airbnb guests said they used Airbnb to explore a specific neighborhood. (Airbnb, 2013b)

This preference for peripheral neighbourhoods seems to contradict our earlier conclusions that location is an important factor in the choice of Airbnb users, and that they tend to behave like regular tourists. Of course, the geographical delimitation of 'central hotel corridor' is quite unspecific. We therefore have to explore where Airbnb visitors concentrate, what impact they have on neighbourhoods, and whether that means a change compared to traditional tourism.

Spatial Concentration

Different international studies consistently find a concentration of Airbnb activity in centric and touristic areas. In Barcelona, the platform's claims were debunked by different studies: 'despite the company's mantra, its offer is not spread over the neighbourhoods of the city but highly concentrated in the central neighbourhoods and strongly correlated with the presence of hotels' (Arias-Sans, 2015). Gutiérrez et al. (2017) even find a stronger link between Airbnb presence and tourist attractions – 'proximity to the city centre and the beach, the presence of sightseeing spots in the vicinity, and activities related to leisure, hospitality and entertainment' – than for hotels, which they explain by the platform's greater flexibility to expand in ready-built areas (Gutiérrez et al., 2017: 290). For Spain in general, '73% of places in tourist accommodation in rental houses is concentrated in the touristic centres of cities, opposed to the regulated offer, which is much more distributed (only 42% situated in the most touristic areas)', with the biggest difference between those two percentages for Madrid: 81% and 19%, respectively (EY España, 2015).

A study of Airbnb in 13 Italian cities (Picascia et al., 2017) confirms this concentration, although it distinguishes three 'archetypes' of spatial spreading: the typical historical art city of Florence shows a growth from the geographic centre (within the historical walls) outwards, whereas Milan, comparable to London and Berlin according to the authors, starts its growth from gentrified areas with a relatively blank spot in the old city. Rome shows different points of origin from which Airbnb has developed, combining tourist hotspots with adjacent residential areas and gentrified neighbourhoods. This study finds, besides an almost perfect but also obvious correlation between Airbnb activity and tourist flows, a high to perfect correlation – if we exclude the outlying cases of Bari and Genoa – between the expansion of urban vacation rentals in local housing stock and the profitability of Airbnb. Some Italian cities show extreme shares of the local rental market being offered as vacation rentals, with 8.0% for Rome, 8.9% for Venice, 17.9% for Florence and up to 25.3% for Matera.

In Chapter 5 we indicated how studies have found performance decays – in terms both of average daily rates and of revenues – from the city centres to the peripheries. The higher demand for centric Airbnb apartments is the driver of its spatial distribution, as Quattrone *et al.* (2016) have shown, for supply growth outside the most popular areas is not met with similar demand growth. Hosts in peripheral neighbourhoods may try to list their residence on Airbnb, but they will rarely find clients.

The spreading pattern can be easily observed by analysing the concentration of Airbnb listings and especially the nights booked per neighbourhood. In most cities, the Airbnb hotspots are unequivocally determined by central location and, as a secondary condition, the character of the district as residential. A good example of Airbnb presence being conditioned by the availability of eligible houses and apartments is the fact that the platform is underrepresented in the City of London but highly concentrated in all the surrounding central boroughs (Figure 7.3). Similarly, the

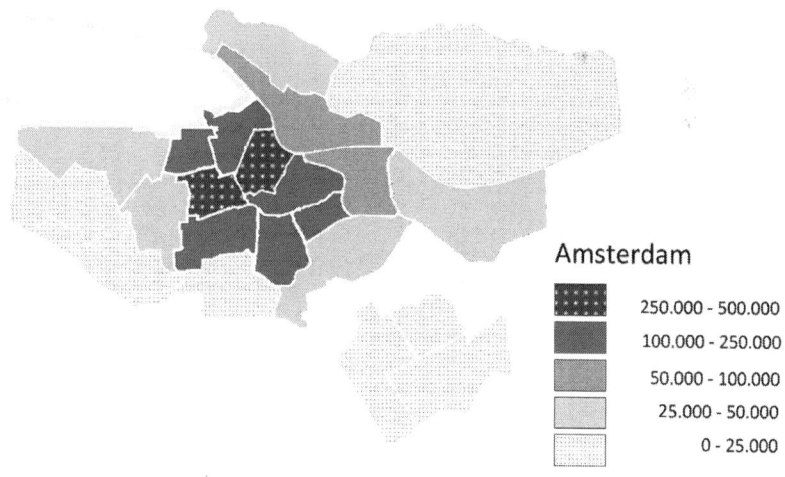

Figure 7.1 Airbnb overnight stays Amsterdam, 2017: (1) De Baarsjes-Oud West, 334,517 (16.1%); (2) Centrum-West, 331,164 (15.9%); (3) De Pijp-Rivierenbuurt, 238,635 (11.5%)
Source: AirDNA, Bakker *et al.* (2018b).

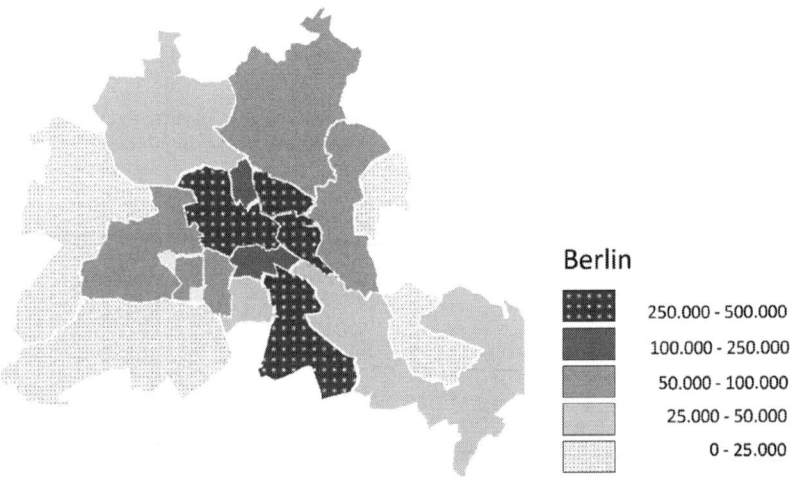

Figure 7.2 Airbnb overnight stays Berlin, 2017: (1) Mitte, 476,119 (22.1%); (2) Prenzlauer Berg, 307,806 (14.3%); (3) Neukölln, 291,516 (13.5%)
Source: AirDNA, Bakker et al. (2018b).

Amsterdam Canal Zone (Centrum West and Centrum Oost) seems to be a favourite, but supply and demand spill over to the neighbourhoods to its west and south, closest to the location of the main museums and the nightlife area (Figure 7.1). Also Berlin has a strong concentration in the city centre and adjacent districts, plus the hip area of Neukölln (Figure 7.2).

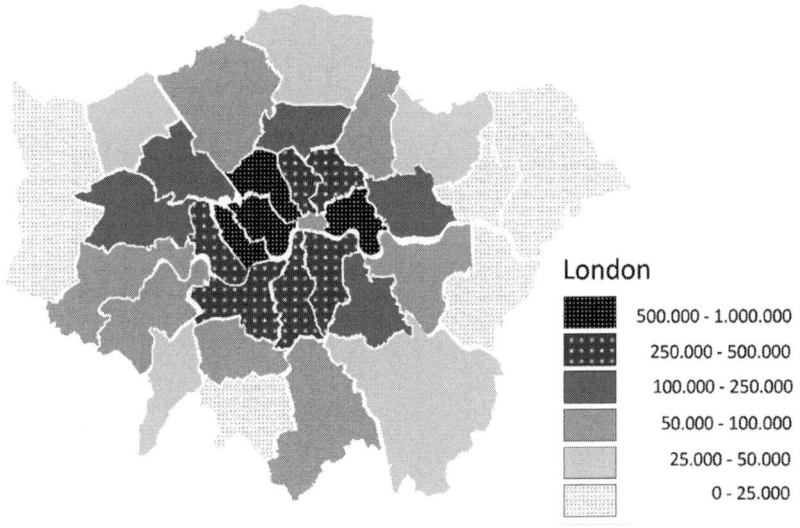

Figure 7.3 Airbnb overnight stays London, 2017: (1) Westminster, 978,537 (14.6%); (2) Tower Hamlets, 758,107 (11.3%); (3) Camden, 582,262 (8.7%)
Source: AirDNA, Bakker et al. (2018b).

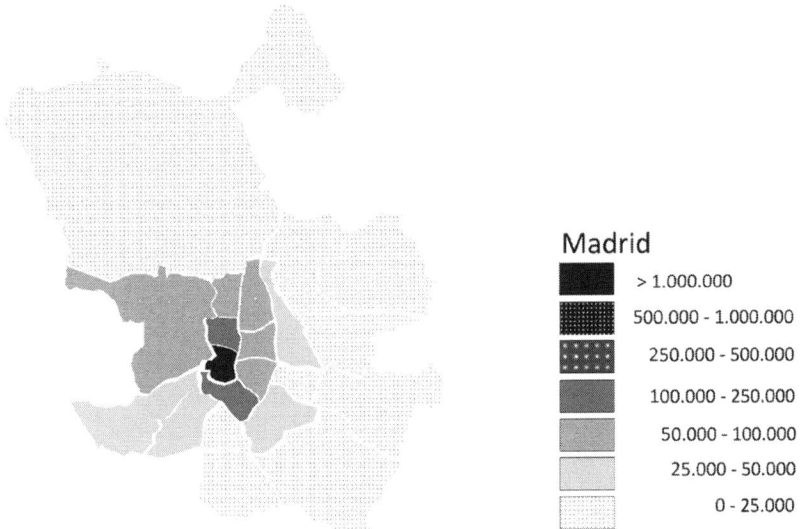

Figure 7.4 Airbnb overnight stays Madrid, 2017: (1) Centro, 1,314,031 (61.0%); (2) Arganzuela, 118,582 (5.5%); (3) Chamberí, 115,011 (5.3%)
Source: AirDNA, Bakker et al. (2018b).

Madrid has an extreme concentration – 61% of all overnight stays – in the Centro district (Figure 7.4). Only Paris shows a more even distribution in a saturated market: Montmartre is Airbnb's stronghold, but the city has no light-shaded areas (Figure 7.5).

A clear link to the availability of residential housing has been demonstrated in Barcelona (Gutiérrez et al., 2017), and can hardly be

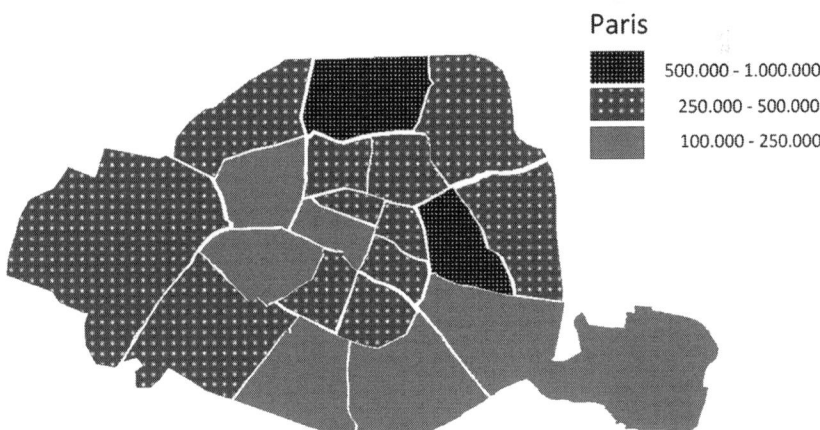

Figure 7.5 Airbnb overnight stays Paris, 2017: (1) Buttes-Montmarte, 645,784 (10.0%); (2) Popincourt, 575,437 (8.9%); (3) Vaugirard, 449,800 (7.0%)
Source: AirDNA, Bakker et al. (2018b).

interpreted as a spreading of tourism to the neighbourhoods, let alone peripheral neighbourhoods as the platform has claimed. Rather, the presence of Airbnb in neighbourhoods adjacent to city centres should be seen as a symptom of gentrification and the expanding 'touristification' of the city. A study of Airbnb emergence in the Dutch mid-sized city of Utrecht (Ioannides *et al.*, 2018) illustrates this spill-over effect to the gentrifying district of Lombok by demonstrating a significant positive relation between Airbnb locations and the presence of services like restaurants, and negative relations for the presence of families with children, low-income households and distance to the city centre. These observations lead to the conclusion that Airbnb supply and demand follow the interest of visitors for 'neo-bohemia' districts outside the tourist bubble, where formal tourist accommodation (traditional hotels) is less available. 'Precisely, because the Airbnb phenomenon is one that responds rapidly to growing demand, it can be an important force of further touristification in such neighbourhoods' (Ioannides *et al.*, 2018: 7).

These factors explain that Airbnb may well be concentrated outside a so-called 'central hotel corridor', but also that this questionable delimitation is not an indication of decentralised location, as suggested. On the contrary, there is a significant distance decay effect from the city centre for both the number of listings and average daily rates. The analysis summarised in Tables 7.1 and 7.2 shows the significant exponential decay of

Table 7.1 Distance decay: Number of Airbnb listings at a distance range from a central point, 2017 (summarised table)

Km	Amsterdam (Leidseplein)	Berlin (Rotes Rathaus)	London (Equestrian Statue of Charles I)	Madrid (Puerta del Sol)	Paris (Hôtel de Ville)
0–1	3691	739	3907	15,878	13,083
1–2	6921	2575	8917	9841	27,919
2–3	8124	3998	16,167	5533	31,873
3–4	4724	6553	18,964	3751	44,753
4–5	1441	9159	28,157	2587	30,488
5–6	597	8929	25,253	1720	7888
6–7	450	10,582	18,409	923	2147
7–8	497	10,889	15,147	813	306
8–9	329	8841	10,783	636	58
9–10	74	2430	7853	207	
10–11	85	1027	7353	186	
11–12	21	764	5172	195	
>12		1832	17,466	114	

Source: AirDNA, Bakker *et al.* (2018b).

Table 7.2 Distance decay of the number of Airbnb listings: ln (Listings) regressed against distance from a central point, 2017

	B	SE B	β	t	P
Amsterdam (Leidseplein)	−0.714	0.020	−0.953	−35.736	0.000
Berlin (Rotes Rathaus)	−0.329	0.007	−0.909	−48.640	0.000
London (Equestrian Statue of Charles I)	−0.337	0.002	−0.985	−151.233	0.000
Madrid (Puerta del Sol)	−0.503	0.014	−0.942	−37.199	0.000
Paris (Hôtel de Ville)	−1.317	0.054	−0.940	−24.410	0.000

Source: AirDNA, Bakker *et al.* (2018b).

the number of Airbnb listings as they are further removed from a centric point in each city.

Gentrification and Overtourism

Following the observations on Airbnb patterns in Milan and Utrecht (Ioannides *et al.*, 2018; Picascia *et al.*, 2017), we see participants – hosts and guests alike – being attracted to areas in transition from lower class to 'hip and happening' neighbourhoods. The processes of gentrification (Glass, 1964; Lees *et al.*, 2013; Smith, 2005), and specifically commercial gentrification (Zukin, 2008, 2009), are closely associated with the emergence of vacation rentals, as the novelty of the area has an appeal for tourists, while at the same time the presence of tourists contributes to the changes in the area. In other words, we cannot solely blame tourism as a cause of these change processes; it is rather that, simply put, tourists share the same middle-class characteristics and taste as the (relatively) local gentrifiers. The imagery of the 'Experience your destination with locals' videos (see Chapter 3) illustrates these preferences. This also places the 'off the beaten track' mantra in a clearer light: Airbnb guests do not just look for any local authenticity, but specifically go after the 'new' or gentrified local authenticity.

This means that, even though the arrival of tourists is sometimes the most visible manifestation of the transformation of residential neighbourhoods, it is not its only symptom, let alone its cause (Blickhan *et al.*, 2014; Füller & Michel, 2014). It is even questionable whether gentrification and 'touristification' develop simultaneously. Through these transformation processes, in fact three groups compete for the use and 'ownership' of a neighbourhood: original residents, gentrifying newcomers and visitors. Usually, a first wave of students, artists and other 'bohemia' followed by the more affluent middle classes displace low-income residents through economic mechanisms. The new residents subsequently start claiming ownership of the neighbourhood's 'authenticity'. The tension that arises when tourists arrive therefore reflects a conflict with multiple sociocultural and economic interests (Gravari-Barbas & Jacquot, 2017; Novy & Colomb, 2017).

It is impossible to objectively establish when tourist pressure becomes excessive. While the economic impact or the environmental footprint of tourism is measurable, it is much harder to determine the sociocultural *carrying capacity* of tourist destinations, as the acceptance of visitors depends on different factors such as cultural affinity and perceived economic benefits for residents. Growing tourism weariness is therefore not only caused by increasing tourist arrivals, but is also mediated by the gentrification process itself: middle-class residents are less likely to be directly dependent on tourism revenues or employment, meaning that the cost they experience as caused by tourism outweighs the benefits. Also, the perception of tourism shifts over time: both Berlin in the 1990s and Amsterdam during the 2008 financial crisis gambled on attracting visitors to alleviate the economic downturn, but may have created a monster that angers their residents in better times (Novy, 2017).

The clearest manifestation of tourist nuisance occurs when visitors and residents compete for the use of available assets. In its simplest form, the groups compete for the same space, leading to complaints about crowdedness. Venice has become intransitable during the summer months. Day visits during the Christmas holiday obliged Madrid's local government to introduce one-way traffic for pedestrian areas in 2017. Tourist masses climbing the hill to the Hollywood sign risk making the surrounding residential neighbourhoods unreachable for emergency services. The victims of the 2017 terrorist attacks in Barcelona were nearly all visitors to the city; locals have stopped going to the Rambla.

This direct competition for parking spaces, restaurant tables and access to museums is accompanied by more permanent transformations. Displacement of services is one of the nuisances for residents: local stores are substituted with souvenir shops and services aimed exclusively at tourists. A striking phenomenon in Amsterdam is the Nutella stores: a relaxed licence regime for ice-cream parlours made this type of outlet a favourite of entrepreneurs tapping into the tourism boom. In general, reduced supply and additional tourist demand drives up the cost of living, not only for groceries, bars and restaurants, but more seriously for housing as well. Normal rents have become unaffordable for schoolteachers or medical staff on tourist islands such as Santorini or Ibiza, worsening the vicious cycle of the visitor economy displacing services required by permanent residents.

Tourist counts in traditional statistics – based on hotel overnight stays – do not accurately represent real visitor pressure, as those who stay with family and friends, in alternative accommodation, and day visitors as well as cruise ship passengers are not included. The latter two categories play an important role, for instance, in Venice, where important numbers of tourists stay in the surrounding cities. The city has 2.6 million hotel guests per year, slightly more than 10% of the total visitors it

Table 7.3 Top 10 European cruise ports

	2016	Passengers	2000	Passengers
1	Barcelona	2.7 million	Cyprus	0.8 million
2	Civitavecchia	2.3 million	Balearic Islands	0.6 million
3	Balearic Islands	2 million	Barcelona	0.6 million
4	Venice	1.6 million	Piraeus	0.5 million
5	Marseille	1.6 million	Istanbul	0.4 million
6	Naples	1.3 million	Genoa	0.4 million
7	Piraeus	1.1 million	Naples	0.4 million
8	Genoa	1 million	Civitavecchia	0.4 million
9	Savona	0.9 million	Venice	0.33 million
10	Tenerife	0.9 million	French Riviera	0.3 million

Source: MedCruise.

receives. Visitor numbers from cruise ships in the Mediterranean have increased dramatically from 2000 to 2016, as shown in Table 7.3.

Airbnb visitor numbers are not made available, except for occasional triumphant press releases by the platform. With scraped AirDNA data, we can base an estimate of yearly individual visitors to a destination by dividing total nights booked by the calculated average length of stay for the destination – usually around 4.0 – and multiplying it by an estimated party size; see Table 7.4. This party size estimate is derived from the

Table 7.4 Estimates of visitor numbers using Airbnb, 2016–2017

	Visitors 2016	Reserved nights 2017 (a)	Length of stay 2017 (b)	Party Size 2017 (c)	Visitors 2017 ((a*c)/b)	Growth 2016–2017
Amsterdam	1,130,000	2,080,000	4.05	2.46–2.91	1,380,000	22.1%
Barcelona	1,880,000	3,010,000	4.01	2.77–3.54	2,370,000	26.0%
Berlin	1,040,000	2,160,000	4.55	2.50–3.00	1,300,000	25.8%
Brussels	420,000	700,000	3.48	2.64–3.29	600,000	44.5%
Copenhagen	890,000	1,780,000	4.52	2.61–3.22	1,150,000	28.8%
Dublin	580,000	1,020,000	3.40	2.70–3.40	920,000	58.0%
Helsinki	150,000	340,000	3.86	2.66–3.32	270,000	72.6%
London	3,130,000	6,700,000	4.21	2.71–3.42	4,880,000	56.0%
Madrid	1,060,000	2,160,000	3.52	2.79–3.57	1,950,000	83.0%
Paris	3,150,000	6,450,000	4.42	2.57–3.14	4,160,000	31.8%
Prague	1,140,000	1,790,000	3.31	3.08–4.16	1,960,000	71.9%
Stockholm	220,000	400,000	4.55	2.52–3.05	250,000	14.2%

Source: AirDNA, Bakker *et al.* (2018b).

accommodation's capacity, as a maximum, and the average of 2.0 and the capacity, as a more conservative estimate. This gives a total visitor bandwidth based on an approximate party size of 2.5–3.0.

Protests against overtourism in cities such as Barcelona, Amsterdam and Reykjavík often associate the phenomenon with the presence of urban vacation rentals. Besides their simultaneous emergence in the last decade, Airbnb and other platforms also generate the most visible interference of tourist activity in residents' lives. As indicated above, however, Airbnb can at most be seen as a catalyst of excessive tourist pressure – as the market leader in low-cost accommodation – but not as its cause: the growth of tourism is due, simply put, to wealth increases in the global middle classes, accelerated by the rise of low-cost travel. But measures to restrict the externalities of tourist pressure often explicitly address the concurrent problems of vacation rentals: new regulations have sought to reduce the supply of vacation rentals, as well as visibly acting, with exemplary fines, against illegal operators.

Further measures against overtourism vary from access restrictions, to economic measures and reconducting tourist streams. Access restrictions have especially been applied in small destinations: Santorini does not allow more than 8000 cruise visitors per day; Cinque Terre in Italy has restricted access to 1.5 million tourists per year – down from 2.5 million in previous years; and the Spanish beach of Playa de las Catedrales requires online reservations, limiting crowds to 4812 per day. Hotel investment stops in Barcelona and Amsterdam seek to contain tourism growth. Demand can also be reduced by price measures: authorities have employed instruments such as tourist tax or value-added tax to limit the use of regulated accommodation. Finally, cities have tried to spread tourists by actively promoting unknown neighbourhoods or surrounding areas. All these examples illustrate the main problem caused by urban vacation rentals: their lack of transparency impedes effective regulation and law enforcement. Airbnb guests are not included in tourist statistics. If hotels become scarce or expensive, the platforms absorb existing demand. And, as we have seen before, Airbnb undermines spreading policies by concentrating tourists in centric areas.

Urban Vacation Rentals and the Housing Market

Impact on the housing market is possibly the most debated external effect of Airbnb and other platforms. Economics textbook principles dictate that, if vacation rentals generate additional demand for residential housing, there must be price effects. Such rent inflation has been hard to quantify, though, as Airbnb has been just one of the factors in the post-crisis demand increase.

Housing market concerns were one of the motives for the New York State Attorney to subpoena Airbnb for data in 2013. The study found 4600 apartments used as vacation rentals for more than three months in 2013, and 2000 for more than half of the year, leading to the conclusion that these units had been made unavailable for the residential housing market. This conversion of long-term rentals into short-term – vacation – rentals had been most frequent in gentrifying neighbourhoods in lower Manhattan and Brooklyn (New York State Attorney General, 2014).

Early studies reported that the relatively small numbers of Airbnb units, compared to the total available housing stock, led to minimal rent increases. Based on 2012 data for New York, a consultancy report (Rosen *et al.*, 2013) stated that this impact would be minimal, although 'admittedly difficult to quantify' – compared to macro-economic drivers such as limited supply and increased post-crisis demand (Rosen *et al.*, 2013: 3). A study commissioned by Airbnb (Kusisto, 2015) found the price increase to be US$6 per month for New York and US$19 for San Francisco, assuming that the vast majority of listings were not taken off the market as they were offered by single-listers; however, even under the assumption that all Airbnb listings would be offered for the entire year by investors, the price effect would remain limited to US$24 and US$76, respectively.

Responding to the estimates in this report, Lee (2016) argues that in the most popular neighbourhoods of Los Angeles, with an apartment vacancy rate of 3%, now 3.5% of the housing stock has been converted into tourist accommodation; in Venice Beach this even amounts to 12.5%. His conservative estimate is that each 1% decrease in supply leads to a 0.2% rent increase. Besides the supply loss, a further price effect is caused by the price surge of regular rents competing with tourist prices. In popular Airbnb neighbourhoods, rents have risen by 16%, compared to a 12% median increase for the entire city (Samaan, 2015), although this difference, as we have already indicated, does not prove causation of price hikes by short-term rentals.

A further nuance is provided in a report commissioned by non-profit organisations that work to preserve affordable housing in New York (Delgado-Medrano & Lyon, 2016). The study detects a total of 8058 impact listings – with strong commercial activity – or 16% of total Airbnb listings for the city. Even if these numbers constitute a small share of total housing stock, their impact should not be underestimated: if these 8058 impact listings were to be added to the rental market, the number of vacant rental units for New York City would increase by 10% and the vacancy rate would rise to 4.0%. This effect is again stronger for popular neighbourhoods: adding the 687 impact listings for West Village/Greenwich Village/SoHo to the housing market would raise their vacancy rate from 2.9% to 5.0%, and the 1157 impact listings for Chelsea/Hell's Kitchen would bring the vacancy rate from 4.2% to 5.7%. While the upward effect on rents may still seem small, the additional demand generated by the platforms prevents prices from going down if other market conditions change (Edwards, 2016).

An analysis of AirDNA data by the McGill University Urban Politics and Governance research group (Wachsmuth *et al.*, 2018) estimated the removal of New York housing from the long-term rental market based on the days per year Airbnb units were available and actually booked. In New York City, 5600 entire home listings, available for over 240 days and rented for more than 120 days, were considered as lost for the housing market, with 12,200 further entire homes, available for more than 120 days and rented for more than 60, at minimum at high risk of being removed. To these listings, 1200 'ghost hotels' should be added: clusters of private rooms offered in the same building that are jointly operated as hotels. Combining both numbers, the total of units lost to the residential housing market ranges between 7000 and 13,500. The direct impact of Airbnb activity on rents is estimated at 1.4% or a yearly US$384, with yearly increases of close to US$800 for popular areas such as Chelsea, Clinton or East Midtown.

Further proof of inequality effects are racial differences in Airbnb activity and revenue: white hosts in predominantly black neighbourhoods, where they represent 13.9% of the population, were responsible for 79% of Airbnb hosting activity and for 73.7% of revenues. The study by Murray Cox (2017a) gives the example of Stuyvesant Heights in Central Brooklyn, where the 7.4% white population produces 74.9% of the Airbnb hosts, with the externalities affecting all residents alike. The conclusion of the study is that through the loss of affordable housing Airbnb has become a 'racial gentrification tool'.

For home owners, both positive and negative effects of Airbnb on the values of their properties can be expected: popular areas offer the potential for Airbnb profits, but the negative externalities of tourism may also drive down prices. In Amsterdam, such a twofold effect was found: Airbnb density, measured in 10,000 reviews in a radius of one kilometre in the previous year, resulted in a price increase of 0.42% in house sales, but also in a nuisance increase of 0.29 on a 10-point scale. The positive effect of potential Airbnb revenues was found to outweigh the downward effects of nuisance (Van der Bijl, 2016). In the United States, a yearly 0.49% in house price growth and 0.27% of yearly rent growth can be explained by increasing Airbnb activity. This effect is larger in areas with smaller shares of owner-occupiers, suggesting that in these areas landlords are induced to switch from long-term to short-term rentals (Barron *et al.*, 2018).

Rent inflation and the reduction of housing stock are both mechanisms that aggravate the problem of 'touristification', or displacement of residents by tourists and tourist activities. The increased cost of living causes a vicious cycle in which residents lose essential services and are increasingly surrounded by a tourist economy. On the Greek island of Mykonos, a normal rent of €500 has become unaffordable for a schoolteacher's salary of €600 (Christides, 2018). In nearby Santorini, 'you have teachers who work in schools in the morning and wait on tables at night' (Smith, 2017).

To an important degree, the level and speed of displacement seems determined by the negotiating power of landlords and tenants. In Iceland's capital, Reykjavik, when it was hit hard during the financial crisis, real-estate investors bought up entire apartment blocks. Visitor numbers to Iceland increased sixfold since 2000: from 302,900 to 1,792,201 in 2016. This immense tourist pressure – Reykjavik has a mere 122,000 inhabitants – has seduced property owners into turning long-term into short-term rentals, resulting in 44% of the rental market being listed on Airbnb (Demurtas, 2017).

With over 130,000 active listings in 2017 – 83,000 of which achieved actual bookings – Paris is Airbnb's top European destination. Displacement of residents has become noticeable, as the population of the 1st Arrondissement fell by 5% between 2009 and 2014, attributed to decreasing birth rates and a profusion of second homes. Short-term rentals are 2.6 times as profitable as long-term rentals – in centric areas even 3.5 times as profitable. Landlords seeking to evict regular, long-term tenants have been found to do so 'through the guise of rehabilitation efforts' (Bowers, 2017: 15).

The rental market in Spain is problematic as real-estate ownership is often speculative and tenants do not enjoy the same degree of legal protection as in other European countries. The Centro district of Madrid has seen rents increase by 36% since 2014. Abusive increases have been reported as a strategy to evict long-time residents. The availability of rental homes has decreased by 7.3% from 2016 to 2017. Displacement also affects home owners, since the negative impacts of tourism – increased cost of living, noise and contamination, departure of friends and neighbours, the neighbourhood being turned into a 'tourist theme park' – are amply compensated for by exploding housing prices, a strong incentive for owners to relocate and to surrender their properties to the 'touristification' process (Aunión & Clemente, 2018; Servihabitat, 2017).

An average rent increase of 25%, with incidental price hikes of 200%, also affects the historic centre of the coastal city of Málaga. The number of residents of the city centre has gone down from 11,000 in 1970, to 5900 in 2007 and 4942 in 2017. Meanwhile, city authorities have been approving between 50 and 70 licences for tourist short-term rentals per month in the same neighbourhood. Besides de facto evictions spurred by rent increases, a new type of rental agreement has been introduced in the city: the discontinuous contract. The exclusion of the holiday months July, August and September for residential tenants brings us closer to the most pessimistic scenarios for Airbnb impact: that of residential neighbourhoods turned into timeshares (Castillo, 2017a, 2017b; Oskam & Boswijk, 2016).

> **Summary: The Commodification of Neighbourhoods**
>
> Urban vacation rentals are not the cause of gentrification and overtourism, but they are a symptom and a catalyst of these processes. They are a threat to affordable housing and to the availability of services required for permanent residents. The displacement of residents and services has started a vicious cycle which transforms neighbourhoods from living environments into a commercial offer of accommodation and leisure facilities for tourist consumption.

Note

(1) 'Basta pegar una ullada al nostre voltant per veure com Ciutat i qui l'habitem ens estem transformant a la força. El centre històric s'està tornant un gueto de guiris d'on les llonguetes en som desplaçades, les tendes de productes bàsics estan desapareixent per ser ocupades per restaurants hipsters, gelateries i botigues de souvenirs, els espais comuns com les places estan sent envaïts per terrasses, els lloguers i el menjar pugen a un preu desorbitat només apte per a la butxaca del visitant adinerat, els nostres carrers, convertits en vies de trànsit pel consumidor-turista, sofreixen la massificació i la desmesurada pressió humana que fa insofrible la vida quan comença l'estiu o quan davallen del creuer ... I enguany encara es preveu xifra record en arribada de turistes!'

8 The Regulation of Urban Vacation Rentals: Empowered Residents, Emasculated Authorities?

> The Sharing Economy is a movement: it is a movement for deregulation. Major financial institutions and influential venture capital funds are seizing an opportunity to challenge rules made by democratic city governments around the world, and to reshape cities in their own interests. It's not about building an alternative to a corporate-driven market economy, it's about extending the deregulated free market into new areas of our lives.
> Slee, 2015: 24–25

Urban vacation rentals constitute in many places an overt violation of an array of regulations aimed at protecting consumers, residents and commercial interests. Advocates of platforms such as Airbnb have argued that the innovations they have introduced make existing legislation obsolete. Authorities should adapt their laws to the emergent phenomenon, or preferably even reach an amicable agreement with the platform, since it contributes to their tourist economies and helps ordinary citizens reap the fruits of visitor streams.

Airbnb, in particular, has responded to regulation pressures by pointing out the property rights of home owners, allegedly entitling them to do as they please with their houses, and their rights to privacy, which justifies denying authorities information about their activities. Besides, as Airbnb CEO Brian Chesky argues, they are perfectly equipped to self-regulate:

> [City-level rules are] primarily set up for screening. To protect consumers. Well it turns out that cities can't screen as well as technologies can screen. Companies have these magical things called reputation systems ... We think that government should exist as the place of last recourse. (Airbnb CEO Brian Chesky, quoted by Slee, 2015: 84)

In this debate, the platform has adopted the neoliberal discourse of the lone individual – backed by Airbnb – who not only faces the power of big business, but also the arbitrary tyranny of a government that tries to supress citizens' entrepreneurial spirit.

Cities around the world have struggled with the responsibility to become informed about their economic development, to reduce the externalities of urban vacation rentals, to ensure the safety of residents and visitors and to prevent crime, to manage tourist masses and to weigh the interests of the visitor economy vis-à-vis the housing market. The ineffectiveness of most measures taken since Airbnb started in 2008 has been an important factor in the growth of its business: similar to the case of Uber, it could be said that part of the platform's business model is 'predicated on lawbreaking' (Edelman, 2017). In addition to a flawed understanding of the nature and impact of urban vacation rentals, different factors have contributed to the ineffectiveness of regulatory actions.

Fragmentation

If we qualify the growth of urban vacation rentals as 'exponential', that also means that it was slow initially. It is understandable that regulatory initiatives were too limited and too late during the first years of Airbnb expansion, especially if we consider that in those crisis years tourist numbers were seen as a life raft for destinations' economies. Understanding and reacting to vacation rental impact did not transcend local tourism policies; regulating the phenomenon was left to city authorities, whose primary concern was getting their proper share of a growing visitor economy by collecting tourist taxes. The curious result was that any concession by platforms concerning local taxes was celebrated as a resounding victory for the regulators, while many higher impositions – income tax, sales or value-added tax, corporate taxes – owed to national or federal authorities were not even contemplated.

An important limitation of such tourist tax agreements, which Airbnb reached with cities such as Amsterdam, Paris, Lisbon and numerous cities throughout the United States (Airbnb Citizen, 2017), is that the platform helpfully assists cities as a tax collection intermediary. Collected taxes are then transferred to the cities in quarterly or annual bulk sums. The advantage for the platform is that this role allows it to protect the privacy of its hosts – to conceal their identities – thus making it impossible for authorities to verify declared quantities or to enforce any further regulations. Periodical payments therefore seem like voluntary donations in exchange for permissive policies, rather than regular tax payments.

Currently, Asian countries in particular tend to develop urban vacation rental regulation at a national level. The current stage of development of the platforms and differences in political culture, as well as the fact that visitor nuisance especially affects affluent citizens in urban condominiums, are perhaps some explanations of this more centralised approach.

Reasons to Regulate

Besides the direct economic benefits of receiving more visitors, it can be argued that platform companies – peer-to-peer (P2P), 'sharing' or whatever they are called – promise further productive innovations for urban economies. Subjecting them to pre-internet legislation would stifle these innovations and thus deprive cities of potential benefits (Sundararajan, 2014). An example would be the French anti-Uber law that forbids GPS tracking of taxis and obliges vehicles to return to their base between rides, measures that reduce the efficiency of their service (Edelman & Geradin, 2015). In many legal systems, property laws entitle home owners to sell or rent their properties; occupant rights can be transferred to tenants, although subletting is frequently forbidden by contract.

Edelman and Geradin (2015) indicate different reasons to restrict these ownership rights, the most important of which are externalities caused by commercial activities. If a home owner exercises his or her right to rent out an apartment on a vacation rental platform, that transaction affects neighbours who have had no part in the contract. Common areas will be more intensively used, increased tourist activity will increase traffic in their streets, it may attract criminal activity, and all these elements may decrease their comfort and wellbeing, as well as the value of their properties. The home owner's decision may have further consequences as described in the previous chapter: it may alter the character of the neighbourhood and reduce the availability of housing, and the existence of vacant homes may affect safety.

In other words, urban vacation rentals produce externalities, the cost of which are currently borne by non-participants. In the failure of regulatory measures, Espinosa (2016) suggests that the damage caused to home owners can also be addressed by common law. Using the Supreme Court ruling in the Metro-Goldwyn-Meyer case against video-sharing site Grokster, urban vacation rental platforms can be held liable for private nuisance; in case such a class action prevails, 'the defendant home-sharing website will be forced to account for the negative externalities caused by the industry, or cease to do business in the manner in which they currently operate' (Espinosa, 2016: 627).

A second reason to apply regulatory measures is information asymmetry between both contract parties. Regulatory frameworks are in place to guarantee quality and safety standards, a function that platforms have replaced by the trust generated in mutual review systems. Besides the fact that these review systems are flawed – as argued in Chapter 4 – it is unlikely that they contemplate maintenance standards whose evaluation requires technical expertise. For example, a guest review will not detect a poorly maintained heating system that leaks toxic carbon monoxide.

In the third place, regulatory intervention may be desirable because customers, despite receiving accurate information, are unable to properly assess risk and take the necessary precautions. Such a cognitive bias could lead a vacation rental guest to focus on salient risks, and therefore avoid neighbourhoods with a high crime rate, and underestimate the benefits of safety measures such as sprinklers, fire exits or a deadbolt on the door. Finally, vacation rentals must be regulated to guarantee a universal provision of service. An Airbnb reservation request is accompanied by a profile picture of the aspirant guest and can be rejected by the host, a situation that – as mentioned in Chapter 5 – has facilitated discriminatory practices. Hotels are required to offer a certain number of accessible rooms and average the additional costs out over all guests, whereas a wheelchair user booking an Airbnb unit adapted to his or her needs would individually bear the additional costs of those provisions (Edelman & Geradin, 2015).

It is also true that in most places regulations are already in place to prevent these negative impacts. Zoning ordinances restrict commercial activities and their externalities in residential areas (Cohen, 2016). Safety standards and commercial regulations can also be applied to urban vacation rentals. The problem is therefore their frequent non-compliance with existing legislation. The example of wheelchair accessibility shows that compliance with regulations always generates additional costs. The usual reaction of traditional hotel businesses and their lobby is therefore to demand a 'level playing field', i.e. to remove the competitive advantage of unregistered rentals because of their non-compliance with commercial regulations and tax obligations. This would require, first of all, a solution to the undefined status of urban vacation rentals, creating registration systems. Additionally, their activity should be integrated into tourist statistics; safety and security regulations, fiscal obligations, visitor identity verification and labour laws should be applied to providers. Registration should also allow authorities to control the dispersal of urban vacation rentals (HOTREC, 2015).

The opposite risk is *regulatory capture*, meaning that regulations are enforced to the advantage of existing interests with access to and influence in regulatory bodies; the example of French taxi laws annihilating the technological innovations of Uber shows that the protection of incumbent businesses may be contrary to consumer interests and the quality of service (Edelman & Geradin, 2015). Society's benefits derived from technological innovations would therefore justify an adjustment of regulations to the new situation (Cohen, 2016; Edelman & Geradin, 2015; Lines, 2015). In view of the regulatory inadequacy for the platform economy, Sundararajan (2014) supports Bryan Chesky's case for a primarily self-regulatory regime.

Existing Regulation: Acquiescence or Prohibition

In practice, we can distinguish three different attitudes local governments have adopted on the topic of urban vacation rentals:

laissez-faire, a full-out ban or specific regulation (Lines, 2015; Nieuwland, 2017). The first of these approaches is either a deliberate decision to allow short-term rentals to develop, or an involuntary inaction, for instance because of political inertia. In most secondary destinations, the benefits of attracting more tourists outweigh the negative externalities of the platforms. But also major cities such as London, Paris and Amsterdam did little initially to contain the growth of vacation rentals, until they were surprised by triple-digit growth. The laissez-faire approach also has important disadvantages for the development of the vacation rentals themselves, as Lines (2015) argues with the example of Phoenix: hosts risk the arbitrary enforcement of existing rules; health, safety and zoning issues are not addressed, nor are tax evasion or fairness in regulation. As a consequence, the short-term rental market will continue to grow in an 'ever murkier regulatory state' (Lines, 2015: 1172) which will in the end harm both rental hosts and the city.

Elsewhere, governmental inaction may also be motivated by the consideration that vacation rentals should be ruled by existing legislation, and that a full ban is expected as a result of future judicial ruling. Countries such as Thailand or Singapore are examples of the latter approach (Heng & Jo, 2016; Sritama & Katharansiporn, 2018). The explicit decision to prohibit urban vacation rentals altogether is rare. Anaheim, home to Disneyland, CA, changed its licence policy into a full ban after complaints about illegal motels in residential areas (Barragan, 2016). New York, San Francisco and Santa Monica have also imposed prohibitive measures or partial bans. All these regulations have been aggressively fought by Airbnb in lawsuits invoking the First, Fourth or Fourteenth Amendments: respectively, the right to Free Speech in the platform's publication of rental advertisements; the right not to disclose host data; and the protection against liability for the platform itself (Cyber Report, 2016; Host Compliance, 2018; Joo, 2016).

Specific 'Sharing' Regulation

Specific regulation is a compromise that generally seeks to legalise non-professional vacation rentals while ensuring health and safety standards as well as tax compliance. This type of regulation imposes limits on vacation rentals, for instance by allowing only partial rentals (one room, or a percentage of the surface of a house) or their equivalent, only rentals in the presence of the legal resident. Berlin's 2016 law on vacation rentals made an exception for cases where the resident would still occupy at least 50% of the house (Loy, 2016). Amsterdam limits the number of guests to four. In New York and Amsterdam, regulations apply to entire homes but not to private or shared rooms. This exception also offers a workaround for hosts: analyses in both cities have shown entire homes to have been recategorised as private rooms, probably without further changes to the listings (Cox, 2016; Kloosterboer, 2017).

Furthermore, the most common maximisation is the number of rental days. A 90-, 60- or 30-day cap, as we have seen in different cities, reduces the efficiency of urban vacation rentals and thus prevents the exclusive use of urban housing as tourist accommodation. As we have argued in Chapter 1, a 30-day cap makes particular sense as this would allow the actual 'sharing' of homes during a holiday absence. Cities such as Singapore and New York apply an inverse rental cap, minimising the duration to three months or 30 days, respectively, either in order to prevent the nuisance caused by transient tourist traffic or the fraudulent use of vacation rentals to keep apartments off the housing market (Channel News Asia, 2017; Host Compliance, 2018; Mahmud, 2017).

The latest restrictions refer to the location of vacation rentals. In May 2018, Amsterdam's new local government announced a full ban of vacation rentals from the city centre (AT5, 2018; Bouma & Middel, 2018). Simultaneously, Palma de Mallorca announced a restriction of vacation rentals to stand-alone houses – forbidding apartment rentals – and only if these are not 'on protected land, within a region around the airport, or buildings not designated for residential use'. In Madrid, city centre vacation rentals will be required to have 'a designated entrance separate from the one used by Spanish residents', which effectively bans all higher floor apartments from the platforms. Similarly, Valencia has banned new rentals from the centre and has further restricted them to ground- and first-floor apartments: as a travel website stated, 'Valencia doesn't want you renting an Airbnb with a view' (Carey, 2018b, 2018c). In August 2018, Portugal introduced legislation that allows municipalities to designate 'contention areas' in places suffering from tourist pressure: in these areas, the number of rentals can be limited, rental rights are not transferable and it is not allowed for a single host to operate more than seven listings (Diário da República, 2018).[1]

These measures show how more recently city authorities have started to recognise Airbnb's marketing fabrications, in particular those about the decentralising effect of urban vacation rentals. The image of 'ordinary citizens trying to make ends meet' by offering their homes on the platform has also made way for an understanding of the accumulating capital behind the vacation rentals. Now that these rentals are no longer seen as a marginal phenomenon operated by 'ordinary citizens', it has become urgent for cities to solve the information void. They need to know where these rental apartments are, who owns them, who stays there and how much revenue they make. Airbnb and other platforms resist disclosing this information because of the privacy of their hosts – the Fourth Amendment in the United States – but also because of the economic benefits of non-compliance: if taxed and subject to regulations, rentals become less profitable and Airbnb's inventory would probably be decimated.

In Chapter 1 we alluded to the data 'arms race' that has emerged between authorities and platforms. Quattrone *et al.* (2016) propose this data collection approach as a basis for a differentiated and 'algorithmic' regulation which ensures a just and sustainable distribution of vacation rental benefits. Transferable sharing rights could further contribute to spreading the benefits to less privileged neighbourhoods. The practical problem that arises, however, is that collected data so far are not precise enough to locate suppliers in order to check and enforce compliance. Algorithmic regulation would therefore again rely on the openness of platforms, which they are unlikely to give voluntarily. The authors propose that the provided data 'should be sufficiently specific to inform policies, but also fairly vague to protect the privacy and safety of customers' (Quattrone *et al.*, 2016: 8), but this does not solve the information void that renders regulation ineffective.

Cities are therefore obliged to build their own parallel databases of vacation rentals through registration systems. Despite the transient tax collection systems that are in place in many cities, the platforms thus avoid further liability for specific rental transactions. In the end, the purpose for cities is to collect the exact same information as hosts share with the platforms. Their failure to gather these data, even with the threat of hefty sanctions, is proof that withholding information from authorities is of strategic interest to vacation rental providers and to the platforms. The San Francisco Office of Short-Term Rentals had received 1082 registration applications by November 2015, and estimated that 4296 or 79.9% of the unique hosts were non-compliant with the regulations that came into effect earlier that year (Brousseau, 2016). In our study of Airbnb development in Reykjavík in 2016, we detected 6047 units pertaining to 3832 hosts (Bakker *et al.*, 2017c). In June 2017, fewer than 150 had been officially registered.

Regulation Enforcement

The reason why most urban vacation rental regulations have been little successful in controlling the phenomenon is that authorities have no access to information that would enable them to enforce the rules. Even scraped Airbnb data only have approximate GPS locations, and no street addresses, of listings. For a host unwilling to receive tax authorities, there is little reason to be overly precise when detailing a listing's location. While most regulators depend in practice on neighbourhood complaints, effective law enforcement would require 'seek and destroy' tactics (Lines, 2015), including door-to-door inspections or the identification of buildings and apartments through internet investigations. This labour-intensive detective work is apparently insufficient to deter hosts with soaring profits.

Moreover, as unregulated businesses, hosts have numerous ways to deceive nosy authorities. The simplest ploy is of course instructing guests to identify themselves as friends or distant cousins to concierges and inspectors. Delisting a unit during office hours and putting it back online

Hosted by Sammy

Amsterdam, Netherlands · Joined in June 2011

★ 3655 Reviews ✻ 1 Reference ✿ Verified

Love design, staying active and meet new people!

Languages: **English**

Response rate: **98%**

Response time: **within an hour**

Figure 8.1 'Verified' Sammy may well be a Russian model

in the evening will prevent it from being detected by authorities while keeping it visible for most holiday bookers. As mentioned before, an illegal entire home listing can be disguised as a private room listing. The 'verified ID' system of Airbnb must have loopholes, if we consider the page of the notorious Amsterdam multilister Sammy (Milikowski & Naafs, 2017), who represents several hundred listings and over 3500 guest reviews since 2011 (Figure 8.1). Sammy's profile photo has been taken from a bigger picture that appears on a Russian stock photo website.[2]

In 2017, the 60-day cap in Amsterdam was enforced by self-regulatory restrictions on Airbnb: the hosts were shown a counter with the number of days that were still legally available, and renting beyond the 60-day maximum was made impossible. The easiest workaround for hosts was to use competing platforms; it is also conceivable that hosts made double listings for one unit on Airbnb, although that would have the disadvantage of losing the economic value of guest reviews. Our study (Bakker *et al.*, 2018a) found a compliance for listings newly created in 2017, but an average of 75 rental days for older listings. Besides rentals through competing platforms, the main explanation was reservations made in 2016 for stays in 2017. Airbnb itself has been shown – in addition to the inaccuracies of its marketing narrative – to be equally deceitful. In November 2015, a transparency initiative showed a favourable evolution of its business in New York, downplaying the role of commercial operators. Cox and Slee (2016) discovered an unusual purge of 1000 listings immediately prior to the publication of the numbers.

Berlin provides a good case study for the effectiveness of regulations. To prevent housing market effects, the city adopted *Zweckentfremdung* or 'purpose alienation' laws, i.e. legislation on the use of residential housing for a different purpose. When these went into effect on 1 May 2016, they were considered as among the strictest in Europe, with fines up to €100,000. There is an immediate effect on the total number of listings, as Cox (2017b) observes in his monthly scrapes: from their peak of 17,372 on 4 February, total listings fall 23.3% to 13,328 on 3 May. For

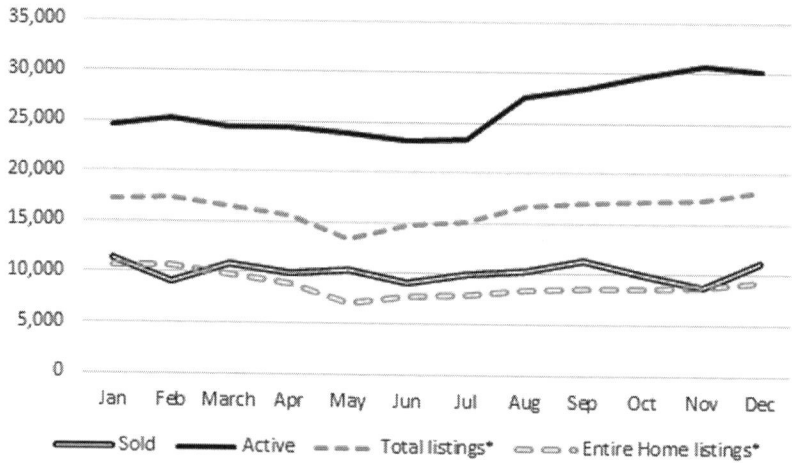

Figure 8.2 Evolution of Airbnb offer in Berlin, 2016
Notes: Scrapes on a single day are lower than monthly totals. Sold: listings with booked days; Active: listings with website activity.
Source: AirDNA, Bakker et al. (2017d), Murray Cox*.

entire homes this effect is even stronger: these listings are reduced 33.7%, from 10,634 to 7064 between February and May. But after this initial scare, both numbers recover: on 8 May of the next year listings are back to a total of 20,576 (+18.4%), and to 10,285 (−3.3%) for entire homes (see Figure 8.2).

AirDNA data, with their algorithmic extrapolation of reservation and revenue data, do not reflect this temporary setback, but they do show a growth slowdown: booked nights in Berlin grew only 68%, little more than half the growth of London or Amsterdam. Before 1 May, entire homes made up around 65% of the offer and private rooms 33%; after the ban, these proportions shifted to 54% and 47%. The approximate division of revenue used to be 80% for entire homes and 20% for private rooms; this shifted to 70% and 30%, as shown in Figure 8.3. In general, we could hypothesise that listings that were active on Airbnb before the rules took effect were not deterred, but that the appeal of the platform to new entrants with illegal entire house listings may have declined.

Two court cases eroded the effect of the regulations. Judges ruled both in the case of a resident who was denied a licence for renting his house up to 182 days, and in the case of second home owners, that their specific activity had no consequences for available housing stock (Redeker Sellner Dahs, 2017; Ting, 2017d; Verwaltungsgericht Berlin, 2016). New rules in 2018 introduced a registration system, allowing licensed hosts to rent out their primary residence for unlimited time and second homes for a maximum of 90 days (Beck, 2018).

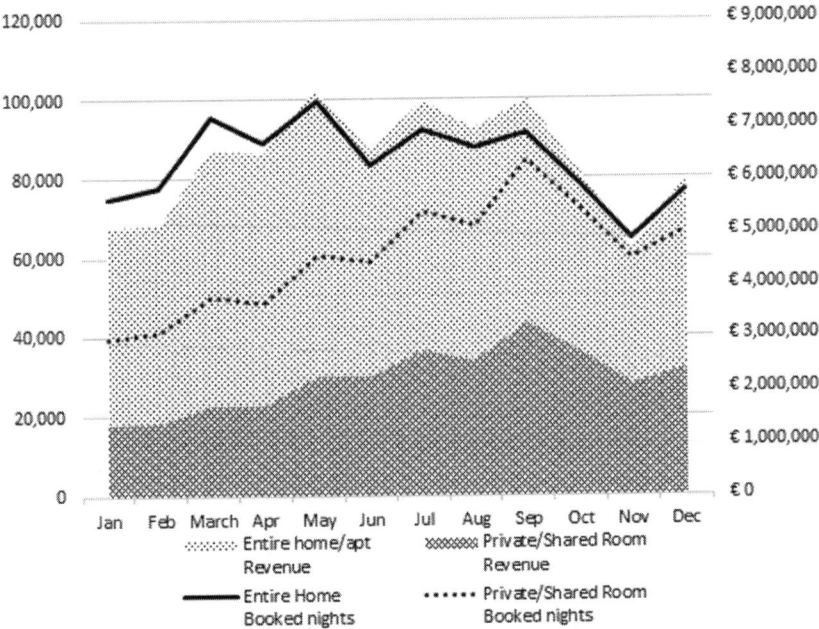

Figure 8.3 Airbnb performance in Berlin, 2016
Source: AirDNA, Bakker *et al.* (2017d).

A Fight for Transparency

It must be clear that transparency is the key issue for cities to control urban vacation rentals and to diminish their externalities. Without transparency, authorities face the impossible task of recomposing the platforms' databases and linking these to their own collection of – truly verified – identity data. Making data available to authorities is required for all regulated commercial activities; however, as indicated, a platform such as Airbnb has invoked the right to privacy of their hosts and their zero liability as intermediaries and has therefore rejected any obligation to disclose commercial host data as unlawful. Remarkably, Airbnb is less principled when it comes to jeopardising access to the Chinese market. It agreed to give up visitor identity and travel data to authorities because, as spokesman Wilczynski explained, 'Like all businesses operating in China, Airbnb China must comply with local laws and regulations' (Bloomberg, 2018).

This respect for China's legal system contrasts with the attitude of Airbnb representatives at home. New York City council member Helen Rosenthal recalls a conversation with Airbnb's Head of Policy, Chris Lehane: 'he said flat out that he did not agree with our laws. […] I did not realize that a $25 billion company can just decide which laws they do and do not agree with' (Griswold, 2015). The context was the debate after New York State subpoenaed Airbnb to turn over user data and found that

72% of the rental activity had to be considered illegal and that 6% of hosts who should be considered commercial – offering at least three listings – collected 37% of all host revenue (New York State Attorney General, 2014). One year later, Airbnb responded with a 'transparency initiative', debunked by Cox and Slee (2016) as described earlier. But besides its lack of truthfulness, the way the platform opened its books as promised in its recent Airbnb Community Compact was a shock to critical politicians and journalists:

> The compact was the entire premise for Airbnb's long-awaited release of data on home-sharing in New York City. What the compact didn't seem to explain was the rigmarole for seeing that data – the appointment, the trip to Civic Hall, the white-walled room with spreadsheets displayed on Airbnb laptops, which visitors were instructed that they could view and take notes on, but not photograph, screenshot, or otherwise copy. (Griswold, 2015)

Recently, however, a few destinations have managed to force the platforms to cooperate after lengthy legal stand-offs. In San Francisco, the platforms were unsuccessful in their appeal against regulation that made them liable for illegal rentals. Negotiations led to an agreement according to which the city's registration system would be incorporated into the rental website, making it compulsory to include the official registration number when posting a listing. The platforms committed themselves to delisting unregistered units. Furthermore, Airbnb and HomeAway broke their usual silence by promising to provide the city with 'a monthly list of all San Francisco listings with sufficient information to allow the city to verify that the unit is in fact registered' (City Attorney of San Francisco, 2017).

Barcelona was the first city to go directly after the platforms, instead of after the hosts. In November 2016, Airbnb and HomeAway were penalised with €600,000 fines for advertising illegal apartment rentals on their sites. For similar reasons, a €300,000 sanction was applied in Mallorca. A Valencian court rejected HomeAway's appeal against a sanction for not including listings' registration numbers, as the platform was found to be a tourist agent rather than just a technical intermediary. Airbnb and local platform Rentalia are awaiting similar appeals. By the end of May, the city of Barcelona announced a new agreement with Airbnb obliging the platform to full openness and compliance: host data – including names, ID numbers and street addresses – were to be shared with the authorities as of 1 June, and unregistered tourist apartments would be delisted (Hosteltur, 2018a, 2018b; La Vanguardia, 2016; Romero, 2017).

This reluctant cooperation is becoming more common internationally and may start to severely hamper the growth of the platform. Japan's restrictive regulations, which also require a registration, led Airbnb to remove 62,000 Japanese listings, 80% of its offer, reducing it to 13,800 in June 2018. Paris has been fining Airbnb US$1200–6200 per unit per day for listings that do not have a tax registration number. The platform paid US$1.6m in fines in 2017 and US$603,000 in the first quarter of 2018. A court case is currently pending which may force Airbnb to remove 43,000 of these delinquent listings, more than half of its active inventory in the city (Carey, 2018d, 2018e).

> **Summary: Compliance and Transparency Stifle Growth**
>
> Urban vacation rentals have benefited from strong competitive advantages as their non-compliance with existing regulations has reduced operating costs. Their growth in residential areas has produced externalities that damage neighbours, in the extent to which they can enjoy their homes or use public spaces, and in the value of their properties. Consumers do not have the same level of protection as in other commercial services, exposing them to health and safety hazards as well as to discriminatory practices. The purpose of local or national regulations is precisely to prevent these abuses. Regulation of urban vacation rentals is only effective if it annihilates their secret weapon and forces them into openness. Registration systems combined with laws that hold platforms liable for illegal practices have recently appeared to bring the growth of the phenomenon back under control.

Notes

(1) Thanks to Dr Marco Martins for sharing this information.
(2) 'Man with spiral notepad', https://ru.depositphotos.com/167537222/stock-photo-man-with-spiral-notepad.html (accessed 25 May 2018).

9 What if This Does Not Stop? Amsterdam 2025

> I live on the fourth floor, there are three rooms here, until recently they were all just rented out. One girl recently left and that room has been put on Airbnb. Totally unexpected, we just arrived one afternoon with our groceries and suddenly we had a bunch of strangers in our house.
> We thought, who are you, what are you doing here; they spoke Italian. We were of course totally surprised, you don't expect this, it is a really strange situation, [...] the landlord didn't discuss that with us, she just did that one day.
> Yes, it causes all kinds of annoying situations, because as a tourist you are on holiday, you come home in the middle of the night. They are in the bathroom when you have to go to work in the morning, and I could continue ... Every time again you have strangers in your house, you don't know who they are, if you can trust them. One day they light up a joint, one day they're throwing up in your toilet. They mess with your stuff; it just affects the pleasure of living here.[1]
> Richard, interviewed by VPRO Tegenlicht (Lubbe Bakker, 2016)

In the aftermath of the financial crisis of 2008–2013, the ideas of the sharing economy were welcomed as a remedy against the perverse excesses of an economic system obsessed with growth. Many residents and policy makers in the city of Amsterdam embraced sharing initiatives as steps towards not only a fairer but also a circular economy, controlled by ordinary people at local levels rather than by faceless multinational corporations. Since the 1960s, the city had built up and groomed its image as a tolerant place for free spirits to experiment with progressive and unconventional ideas. ShareNL, a hub for sharing platforms, proclaimed Amsterdam as a Sharing City in February 2015; a corresponding Action Plan (Amsterdam Smart City, 2016) was adopted by local authorities in 2016, making the Dutch capital the first European 'Sharing City'.

This enthusiasm had coincided with a historic low in real-estate prices, a stagnating housing market because of stricter mortgage rules and a slowdown of building activities. The economic importance of tourism had increased during those difficult years, when the city had effectively been marketed internationally. Airbnb and other 'sharing' platforms could flourish under these circumstances. When eventually their growth made it clear that they did not empower poor city residents, but that they instead were another faceless multinational corporation, while at the same time the economic tide had turned and housing had become scarce and unaffordable again, that created the conditions for a 'perfect storm'.

It would be unfair to criticise politicians for an ideological interpretation of the urban vacation rental phenomenon in those early years. The positive interpretation was dominant, and probably accompanied by strong lobbying campaigns. Concerns were mainly voiced by traditional hotel companies. An attempt to reverse the trend could easily have been framed as short-sighted conservatism at the service of big business. Furthermore, academia was affected by a similar short-sightedness, although maybe not the same inertia.

The academic world did not take the responsibility, at least not in time, to analyse the societal impact of these innovations and to identify the risks. It may have failed both in initiating a conceptual discussion on the nature of urban vacation rentals, and in discussing the research findings with a wider audience. A one-sided interest in empirical studies has resulted in a lack of foresight and delayed the debate; the difficulty of accessing reliable data, required for scoring on research metrics, has affected the interest in the topic. But discipline paradigms have also contributed to repetitive approaches to the platforms' business performance. It is hard to understand why academic journals have echoed 'sharing' mantras even in studies that follow the usual business analysis templates. How can it make sense to publish, for instance, recommendations for Airbnb hosts to increase their revenues, without questioning the assumption that these hosts are ordinary residents in peripheral neighbourhoods who lack basic professional preparation and are unlikely to read trade journals, let alone academic literature?

'A Great and Unique Approach'

The combination of a utopian message, a strong lobby, economic necessity and a lack of critical analysis drove the city of Amsterdam and its policy makers into the arms of Airbnb. One of the main concerns was, initially, the payment of tourist taxes. The only imaginable explanation for this myopia was that Airbnb was seen as a local phenomenon: whereas most taxes applicable to businesses are collected by national agencies, this is a local imposition. Also, the price effects of tourist taxes had become an issue for the hotel lobby, which demanded a level playing field for traditional accommodation and vacation rentals.

At the same time, protests by the hotel sector against unfair competition with illegal hotels and complaints by residents about noise and other nuisances had to be addressed. The municipality of Amsterdam reached an agreement with Airbnb that gave the platform a recognised status – an extremely important precedent in its global expansion – in exchange for promises to share aggregated data with the city and to facilitate the communication about regulation with Airbnb hosts. Although not specified in the Memorandum of Understanding (MOU), the platform started collecting tourist taxes from its users which it would hand over to the city. The payable amount was based on the data unilaterally provided by the platform, for which the authorities would have no verification instrument; these payments can therefore be seen as a donation by the platform to the city. The 'sharing city' of Amsterdam thus guaranteed that it would endorse the vacation rental platform: 'The parties agree that they will communicate positively about the cooperation agreed upon in this MOU' (Municipality of Amsterdam/Airbnb, 2014: 3).

While the advocates of 'sharing' considered the cooperation between the city and Airbnb a 'great and unique approach' (ShareNL, 2015), others questioned the asymmetry of the deal, which allowed the platform to maintain its information advantage, or at best made the city complicit in its non-transparent policies. The agreement gave the cities no instruments to enforce their regulations or control the compliance of Airbnb hosts, nor to base its tax assessment on anything other than trust in the company. Since the agreement was entirely local, income taxes, corporate taxes or value-added tax remained unaccounted for. As a journalist critical of Airbnb summarised in a TV debate: 'The Municipality of Amsterdam is always proud when they have closed a deal with Airbnb. And then you know that those guys are high-fiving in their plane back to Silicon Valley, thinking, what kind of morons are these?' (KRO-NCRV, 2017).

After two years, when the agreement proved ineffective in eliminating or reducing the excesses of urban vacation rentals, the city renegotiated a deal with Airbnb in which the platform obliged itself to enforce the city's 60-day rental maximum by including a counter on each unit's website (Municipality of Amsterdam/Airbnb, 2016). Data show that this cap was formally effective, but it is not hard to think of workarounds for Airbnb hosts: list your property as a private room, not affected by the maximum; list your property twice; or list your property on competing platforms such as Wimdu or HomeAway. When despite these measures Airbnb's growth and nuisance continued, the city halved the allowed rental days to 30. This time, Airbnb was less cooperative and attempted to mobilise its host community against the decision. As Airbnb's local lobby spokesperson or 'Public Policy Manager' declared: 'The Airbnb community – consisting of 19,000 Amsterdam landlords – is disappointed in your intention to have large hotels prevail over Amsterdam families who occasionally share their homes and punish them for the shortcomings of other platforms to promote responsible holiday rentals. Airbnb has been a proactive and supportive partner of Amsterdam since 2014' (Lomas, 2018).

Triple-digit Growth

In scenario narratives, extreme projections are sometimes displayed that, even though they seem unrealistic, become plausible because of a future course of events. In the case of Airbnb development in Amsterdam, however, it is past data that seem hard to believe. Spurred by a climate of political acceptance or encouragement and economic tailwinds for tourism growth, Airbnb demand exploded with triple-digit growth in 2015 and 2016, the first years for which exhaustive scraped data are available, as can be seen in Figure 9.1. Remarkably, double-digit growth in 2017 could be interpreted as a sign of a flattening growth curve, or in other words of a market becoming saturated. But if we take into account the dramatic shift in the political and social acceptance of vacation rentals, the importance of this growth should not be underestimated. These numbers contradict claims by policy makers that the measures had taken effect and that Airbnb had ceased to grow.

Is further growth possible? Conditions are, in the first place, a further growth of tourism, which is highly likely in view of UNWTO forecasts, based on demographic and economic factors. In the second place, there must be room for Airbnb supply growth. As housing stock in city centres is limited, this will depend on the continued displacement of residents by visitors. Increasing housing prices are a first driver behind this development. A second factor is the disappearance of services for residents, neighbours and a less well-definable 'local atmosphere' in which residents feel at home or to which they are used. This tends to be a self-accelerating trend as every resident leaving the city aggravates the situation. In the third place, measures to contain the growth of urban vacation rentals must fail. Amsterdam has the conditions for the 'perfect storm' of the presence of these three conditions being highly probable.

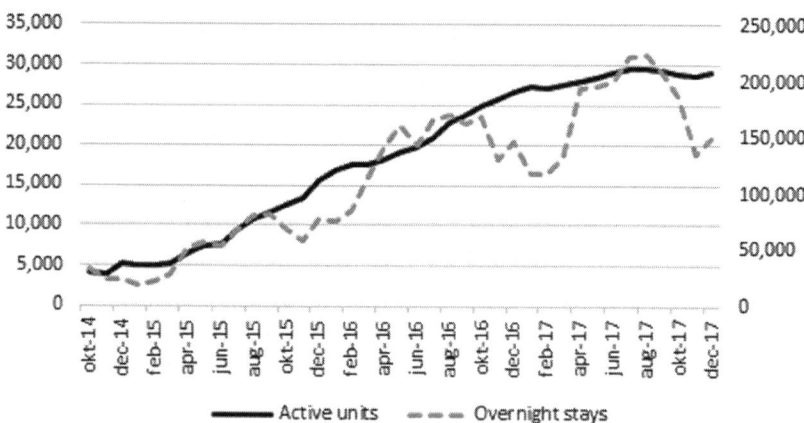

Figure 9.1 Airbnb growth in active units and overnight stays 2014–2017, Amsterdam
Source: AirDNA, Bakker *et al.* (2018a).

Table 9.1 Growth prospect of Airbnb visitors through Airbnb in Amsterdam (×1000)

	2017	2018	2019	2020	2021	2022	2023	2024	2025
Sustained growth (22.1%)	1380	1685	2057	2512	3067	3745	4573	5584	6818
Flattening curve (17.5–10.7– … – 1.9%)	1380	1621	1794	1922	2019	2095	2156	2205	2246
UNWTO growth numbers (3.5–4.5%)	1380	1442	1507	1575	1646	1720	1797	1878	1963
	1380	1428	1478	1530	1584	1639	1696	1755	1816

Source: AirDNA, Bakker *et al.* (2018b).

It is not possible to numerically predict further growth of urban vacation rentals based on past performance because of the uncertainties in this process. To imagine the potential impact, Table 9.1 and Figure 9.2 display the effects of sustained growth, a flattening curve, growth according to UNWTO growth numbers (UNWTO, 2018) and limited growth.

Urban vacation rentals were just one element driving the growth of visitor numbers. The hotel market has been booming; all reports concur in forecasting increasing occupancies and revenue per available room (RevPAR), with Amsterdam clearly outperforming other Dutch cities (Horwath, 2017) and expected to be among the strongest performing cities in Europe (PWC, 2018). The total reported number of visitors – based on hotel overnight stays – amounted to 7,027,000 in 2016 (Fedorova *et al.*, 2017), a growth of 27% since 2012. Equally strong was the presence of both sea and river cruise passengers in the city, with 330,000 and 467,000 arrivals in 2016, respectively (Port of Amsterdam, 2016).

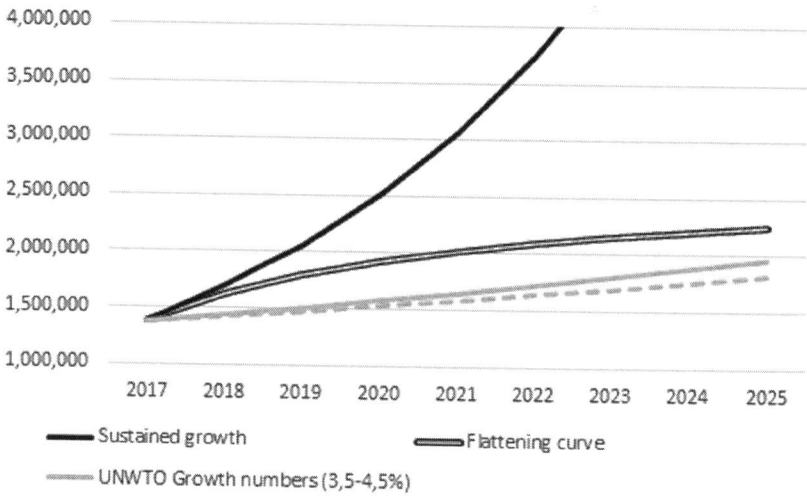

Figure 9.2 Foreseeable growth of visitor numbers through Airbnb, Amsterdam
Source: AirDNA, Bakker *et al.* (2018a).

For a number of years, the residents of centric neighbourhoods had shown increasing signs of tourism weariness. Initially, protests had been dismissed as a refusal to bear the side effects of economic growth by the wealthy middle classes, those who live in the posh 'canal belt'. But with the continued growth of tourism and displacement of residents, this NIMBY discomfort turned out to be an early warning sign of a generalised rejection of the impact of tourism growth. For the local elections of 2018, overtourism had become a crucial topic, with debates around vacation rental regulations, hotel investment stops and cruise tourism policies (Amsterdam in Progress, 2018; Parool, 2018). Venice was a frequently used metaphor in these debates for the black scenario of a depopulated city turned into a tourist theme park.

The 'Venice of the North'

Ironically, the 'Venice of the North' had been one of the marketed nicknames for Amsterdam for international visitors, because of its large number of navigable canals and waterways. What if Amsterdam continues to follow the example of the Italian city's 'touristification'?

The most visible manifestation of excessive tourist pressure is the absolute overcrowdedness – in the sense that streets are full and intransitable – of the city. This condition makes the regular use of city facilities for residents difficult or impractical. As tourist behaviour is more homogeneous than that of residents – they generally want to see the same things in similar timeframes – the situation must be managed with routing systems similar to those used in theme parks:

> Venice's controversial mayor, Luigi Brugnaro, announced that a radical solution to the city's overcrowding problem would be trialled over the bank holiday: the setting up of metal tornelli (turnstiles) at two key entry points: the Calatrava Bridge at Piazzale Roma, where car and coaches arrive, and Lista di Spagna, outside the railway station. The decision was made after the Easter weekend when 125,000 visitors descended on the city on Easter Sunday. The intention is to segregate tourists from locals on the main routes into the city historic centre if incoming visitor numbers become uncontrollable. (Brunton, 2018)

Some of the causes of depopulation are specific to the Italian city – natural circumstances and the resulting housing scarcity, maintenance costs of local housing, accessibility – but warning signs for similar developments can be seen in the situation of Amsterdam. Crowdedness is a subjective, unspecific complaint of city residents which, however, becomes measurable in public transportation capacity problems, especially on the ferries across the river. Also, the public and political debate around Amsterdam's Nutella stores – which popped up like mushrooms after ice-cream vendors achieved a lighter licence regime than bars and retail stores – is a signal of the displacement of local services that was among the drivers of the exodus from Venice. The same goes, finally, for the rising costs and more limited availability of services of shared use for residents and visitors, such as bars, restaurants or cultural institutes.

Amsterdam's economy is much more diversified than that of Venice; nevertheless, these trends may also give tourism the upper hand locally. One of the current specific concerns of the hotel sector is that rising housing prices affect the availability of labour. Affordable housing shortages may not only put other industries depending on low-paid labour at risk, but may also influence the footloose tech companies to whom Amsterdam had a certain appeal in past decades. The availability of student housing will deteriorate dramatically, and go through intermediate stages of young people involuntarily sharing their residence – as we have seen in the opening quote of this chapter – with tourists. The 'discontinuous' rental contracts that are being offered in Málaga – residents can stay from October to May, the summer is for tourist use – echo the darkest scenario of tourism that we described a few years ago:

> Amsterdam had embraced P2P rentals during the financial crisis when it helped home owners afford their mortgages. Airbnb was also seen as a means to increase visitor numbers and to spread tourist spend, but also the nuisance, over the city. The city had always appealed to young and hip visitors looking for experiences off the beaten track. A well-supervised and safe offer of alternative accommodation would underscore that image.
>
> Initially things worked out this way. The abundant availability of non-traditional accommodation also inspired numerous innovative hotel concepts and had spin-offs for other creative businesses. But residential rental prices went up as landlords started to include an 'Airbnb premium' in popular tourist areas. Tourism growth meant an unneglectable investment opportunity for investors, who by purchasing properties started building portfolios of popular Airbnb listings. In 2018 Chinese investor Hui Wang, former owner of a Dutch premier league football club, bought the larger part of the popular Pijp district. New tenants were now offered ten-month contracts. Not unlike a time-share construction, this neighbourhood would thus be entirely available for holiday rentals in the city's peak months. (Oskam & Boswijk, 2016: 34)

An intriguing question is how the processes of gentrification and 'touristification' will interact. Tourists are attracted to the authenticity of the 'off-the-beaten' track neighbourhoods which, in the process of gentrification, start offering facilities for a middle-class public. Generally, despite what gentrifyers and tourists have in common, when visitor numbers grow, this compatibility ceases to exist. Gentrifyers sympathise with those visitors that admire the cityscape in ways similar to themselves, but object to more profane uses by partying holiday-goers, which in their eyes degrade the neighbourhood. The displacement of local stores erodes authenticity and brings globalised facilities and brands that can be found in any other place: 'the streets that used to smell of weed now smell of Belgian waffles' (Pinkster & Boterman, 2017: 466). The reaction of middle-class residents is twofold. In the first place, they use their social capital to voice their discontent to influence tourist policies, and in the second place they consider leaving: either temporarily, as in fleeing the city during the high season or festivals, or as a permanent displacement (Pinkster & Boterman, 2017). This again illustrates the vicious 'touristification' cycle, for every resident that moves out of an attractive neighbourhood creates an opportunity for tourists and tourist businesses to come in.

The city's tourism strategy has been aimed at replacing the young party tourist, attracted by its permissive reputation around alcohol, drugs and sex, with more affluent cultural tourists interested in art museums and architectural heritage. The growth of mass tourism attracts a different kind of visitor, initially mainly travelling in groups and from non-traditional countries of origin. These groups may partially replicate the behaviour of earlier tourists, inspired by the city's image: they will visit the Red Light District and the museums. But their length of stay will shorten and their spending pattern will be different, again benefiting globalised brands and outlets over local specialties. Simply put, Golden Age paintings and cannabis leaves will remain popular but mainly as decoration for T-shirts and coffee mugs.

Since January 2017, the larger part of the city has been subject to a 'hotel stop' that forbids new hotel projects or the expansion of existing hotels in the popular areas (Municipality of Amsterdam, 2017). This has positively affected the values and revenues of existing hotel businesses, but at the same time it has pushed up the numbers of day visitors, either because tourists arrive by cruise ship or because they stay in hotels in the surrounding cities and travel to the Amsterdam tourist hotspots during the day. The main opportunities for further expansion of the booming tourist industry will arise precisely in the neighbourhoods affected by 'touristification': in the houses in attractive neighbourhoods that have been vacated by residents.

In other words, this scenario implies a full convergence of urban vacation rentals with traditional accommodation. City apartments will offer a hotel-like service aimed at the full spectrum of visitors, rather than just the 'off-the-beaten track' segments. The host community professionalises even further, either because residents turn their house into a business and move out of the city, or because investors buy up city real estate. Competitive forces in this market and the privileged position of Airbnb in its negotiations with the city will push rival platforms into a marginal role, with one important exception: Booking will benefit from its strong grip on the traditional hotel market to compile a comprehensive offer at the best locations in the city. The growth of vacation rentals, most of which operate under the radar of local and national regulations, leaves city authorities destitute of instruments to manage tourist streams – other than turnstiles on the canals for crowd management.

> ### Scenario 1: A Platform Complements the Hotel Offer in a 'Touristified' City
>
> In summary, a first scenario for the future of urban vacation rentals is one in which the current booming trend does not stop. The further growth will lead to an oligopolistic control of the apartment market by a limited number of 'winner' platforms. Airbnb – or any other platform that becomes dominant – will converge with the traditional hotel industry. Cities enter into a vicious cycle of displacement of local services, depopulation and loss of regulatory control. A tourist monoculture will cannibalise other economic sectors.

Note

(1) 'Ik woon hier op de vierde verdieping, er zijn hier drie kamers, die werden tot voor kort gewoon allemaal verhuurd. Eén meisje is laatst vertrokken en die kamer is vervolgens als Airbnb te huur gezet. Totaal onverwacht, het was helemaal niet afgesproken, we komen gewoon op een middag thuis met een boodschappentas en er stonden opeens allemaal vreemde mensen in mijn huis.

We dachten, wie ben jij, wat moet jij hier, ze spraken Italiaans, echt totaal verbaasd natuurlijk. Je verwacht het niet, het is gewoon een hele rare situatie [...] de verhuurder heeft het niet met ons overlegd, ze heeft het gewoon van de ene op de andere dag zo gedaan. Ja, het brengt allemaal vervelende situaties met zich mee, want ja, als je toerist bent ga je vakantie vieren, kom je midden in de nacht thuis. Ze staan in je badkamer als je 's morgens naar je werk moet, ja zo kan ik nog even doorgaan. Er zijn iedere keer weer vreemden in je huis, je weet niet wie ze zijn, of je ze kan vertrouwen. De ene keer dan steken ze een joint op, dan weer hangen ze te kokhalzen boven je toilet. Ze zitten ook gewoon tussen je spullen, het is gewoon een flinke aantasting van je woongenot.'

10 What if the City Fights Back? Barcelona 2025

> We are not against tourists nor against tourism, we are also
> tourists ourselves and we know that travelling can be
> a very enriching human activity.
> The problem is not tourism but the current tourist model. A tourist
> model built on a capitalist model that concentrates the benefits in
> very few hands, that causes the destruction of our surroundings,
> that causes rents to rise up to a point where it expels us from our
> neighbourhoods, removing people from their usual networks
> and from their support base, due to which the reproductive
> labour corresponding mainly to women becomes even more
> individualised resulting in higher workloads, a model that
> further downgrades jobs, that turns cities into shop windows
> and that destroys neighbourhoods ... In short, a tourist
> model that generates benefits for very few people
> and that worsens the living conditions for the vast majority.[1]
> Communiqué of anti-tourism activists in Barcelona (Arran, 2017)

Barcelona is perhaps the best showcase of everything that is good and everything that is bad about city tourism. The 1992 Olympic Games can be seen as a celebration of the transformation of Spain into a self-confident democracy, fully integrated in Europe. They enabled the city to completely rehabilitate the degraded port areas and coastline, turning these into attractive residential and leisure zones. An international event such as the Olympics also allows a city to redefine its international image by establishing new traditions and symbols (Moragas, 2014). The appeal of this vibrant and fashionable image brought a balanced increase of visitors: Barcelona not only became an important tourist destination, but also the world's third congress city and a magnet for international students, with close to 10% of the population at the region's universities coming from abroad. The city is in the top 10 in a diverse array of international rankings, ranging from international competitiveness, digitisation, attractiveness, foreign investment, capital of fashion, European tourist numbers and global cruise passenger arrivals (Barcelona Activa, 2017).

Visitor numbers increased fivefold, from 1.8 million arrivals in 1990 to 9.0 million in 2016. The number of hotel beds has increased from 18,569 in 1990 to 67,640 in 2016. In 1990, Barcelona airport had a total of 3.5 million international flight passengers, a number that rose to 32 million in 2016. The city has become Europe's number one cruise port, growing from 115,000 passengers in 1990 to 2.7 million in 2016. One of the most significant shifts can be seen in the reasons to visit the city: whereas in 1990 22.7% came to the city for leisure and 69.1% for business, in 2016 these proportions had shifted to 55.9% and 36.0% (Ajuntament de Barcelona, 2016; Duran, 2005).

Compared to Amsterdam, the Barcelona housing stock had traditionally been less regulated and more privately owned. In 2017 the city had 34,000 Airbnb listings with sales, giving an estimated visitor number of 2.4 million – on top of the approximately 9 million staying in hotels. More than half of these listings were concentrated in the old city centre and in the 19th century Eixample. Two-thirds of the Airbnb units were operated by multilisters, who accounted for 78% of total Airbnb revenue.

'Neighbours: A Species Threatened with Extinction'

Apart from the coastal strip and the Olympic Village, the regeneration process prior to 1992 did not include the rehabilitation of residential housing in the centric low-income areas; instead, it focused on public spaces and cultural infrastructure. This means that, again unlike Amsterdam, neighbourhoods such as the Barri Gòtic, the Raval and the Barceloneta were not primarily gentrified by a more affluent middle class displacing the original working-class population, but rather that the original residents were directly confronted by a tourist 'take-over' of their neighbourhoods. While on average the Barcelona population decreased by 0.6%, the Barceloneta lost 6.6% of its residents – with rents increasing by 9.1% in one year. In the Barri Gòtic, prices went up by 6%, and 17.6% of its residents left (Cócola Gant, 2015; Verdú, 2015).

The displacement mechanism has been analysed by Cócola Gant (2015), in a study based on a survey and interviews with residents of the Barri Gòtic, the Raval and the Born. Housing issues explain one-sixth of the departures, while neighbourhood issues, either by themselves (45.8%) or in combination with housing (27.3%), are behind the majority of these cases: '65% of respondents have lost more than one friend in the area' (Cócola Gant, 2015: 14). A first reason mentioned in the interviews is the loss of commercial services and facilities aimed at residents. This is similar to the process of 'touristification' in Amsterdam, but with changes in central food markets such as the Boqueria – from basic local food supplies to the sale of ready-made food and drink services for tourists – Barcelona offers particularly emblematic examples of commercial gentrification.

A second 'mechanism of exclusion' is affordability. Prices per square metre for new dwellings in Barcelona tripled from 1996 to 2008 (Balanzó & Rodríguez-Planas, 2017). In retail, price hikes were mainly visible in luxury services such as restaurants and ice cream and clothing shops. More importantly, a clash of lifestyles affects local residents through the substitution of traditional bars with European-style restaurants and pubs: 'many of these bars where I used to go became something different. Now they sell "brunch" and things that are not for us. Then you see there are only few places where you can feel comfortable' (Cócola Gant, 2015: 17).

A fourth factor causing displacement is the privatisation of public space – in particular, the expansion of commercial terraces in public squares, to the detriment of public benches. Further nuisances for residents are noise pollution, hygiene and air pollution – for instance caused by restaurants – and 'lack of physical space' or general crowdedness (Cócola Gant, 2015). As mentioned earlier, a very sad example of this displacement was that in the 2017 terrorist attacks to the central Rambla, almost all the victims were visitors to the city; as residents explained, they did not go there anymore.

Anti-tourism Protests

The above means that the gentrification and 'touristification' processes in Barcelona differed substantially from those in Amsterdam. Initially, the cultural transformations and increasing price levels of gentrification were met with discomfort and protest, but these had the counter-weight of economic advantages and neighbourhood rehabilitation (Hernández Cordero, 2015). The discontent grew as the tourist gentrification – meaning displacement of services rather than of residents – expanded to the still mainly residential outer ring of the historic centre: the Raval and the Born (Cócola Gant, 2015). Discomfort turned into anger when tourists invaded and culturally threatened public spaces and houses considered as theirs by residents. Young Italian tourists running naked through the Barceloneta sparked the popular manifestations against tourism and vacation rentals in the summer of 2014 (Baquero, 2014).

The excesses of tourism and tourist gentrification became part of the political debate (Arias-Sans, 2018). Linked to the 15-M (15 March 2011) protest movement against the austerity policies during the financial and economic crisis, and in particular to local housing activism and residents' associations, the left-wing party *Barcelona en Comú* won the municipal elections of 2015. Its programme acknowledged that tourism was an important source of income for the city, but that Barcelona should not become a tourist monoculture. It proposed a range of measures to address the negative effects on housing availability, service displacement and pollution (Barcelona en Comú, 2015). A moratorium on new investments in tourist accommodation – including both traditional hotels and urban vacation rentals – came into effect in 2017.

Urban vacation rentals in Barcelona were considered illegal unless they were formally registered as tourist apartments at the Chamber of Commerce, and hence subject to the regulations applied to hotels or bed & breakfasts. This outlawed the usual practice of Airbnb, marketed – despite evidence to the contrary – as occasional rentals by non-professionals. Also in 2015, the city authorities started acting against illegal tourist apartments with door-to-door inspections, apparently executed by external companies normally active in market research. Violations were punishable with fines of up to €90,000, with an average of €15,000, 80% of which would be pardoned if the owner agreed to offer the apartment as social housing for at least three years (El Mundo, 2015; Soriano, 2015).

The summer of 2017 saw new outbursts of anti-tourism protests, with radical youth organisation Arran damaging tourist buses and rental bikes in Barcelona, and continuing their protests several days later by invading a marina and a local restaurant in Palma de Mallorca. This showed, in the first place, that any effect political measures might have was overshadowed by the continued growth of Spanish tourism, more successful than ever in Europe's economic upswing after the crisis, combined with high insecurity in other Mediterranean beach destinations. In the second place, it was a sign that the debate had become ideological and that tourism was no longer just criticised for its direct externalities, but as a more general manifestation of social injustice (Huete & Mantecón, 2018).

Political Turmoil

In the autumn of 2017, Barcelona experienced political unrest as the result of a conflict between Catalonian secessionists and the central Spanish government. From 2016, the Catalonian region was governed by a coalition of conservative, republican and radical left-wing advocates of independence. This situation is relevant to our topic for two reasons. First, it explains the ideological polarisation that had become manifest in the debate around tourism, despite the fact that conservative Catalonian nationalists had identified tourism as a cornerstone of economic stability should the country become independent (Sala i Martin, 1998, 2001). Secondly, it is relevant because the turmoil led to a 15–20% decline in tourism during the last months of 2017 (El País, 2018).

It is usual to interpret this decline as a reaction of leisure travellers to political insecurity. In the case of Catalonia, just as in the United States, the change in government apparently deterred visitors even if it did not lead to a noticeable increase in public unsafety.[2] As a hypothesis, decreased visitor interest may be the result of a deteriorated image rather than of safety concerns: the strong manifestation of nationalism in cases like these may negatively affect the brand image of tourist destinations as places open to welcoming foreign visitors.

Barcelona, 'the Pebble in Airbnb's Shoe'[3]

Legally, Barcelona treats Airbnb hosts as commercial accommodation providers, which means that the offer is illegal unless it is registered. Unlike other cities, local authorities act not only against individual hosts but against the platforms themselves for advertising illegal accommodation: Airbnb and HomeAway were fined €600,000 each in November 2016, and Airbnb was threatened with another fine of the same amount in the summer of 2017 (El Economista, 2017; La Vanguardia, 2016). The platform expressed its disappointment about being treated according to existing laws, rather than being able to provoke the establishment of new regulations that it considered more appropriate for its business model: in the negotiations, it had demanded the acknowledgement of the figure of the 'home sharer'. With the platform's proactive attitude and its contribution to the local economy, Barcelona should have given in to the pressure but, as the Spanish director of Airbnb stated: 'The Municipality prefers conflict over agreement.'

The position of Airbnb in the debate is that their 'sharing' business creates a new reality that cannot be regulated with old and 'confusing' legislation. Therefore, tailor-made agreements should be negotiated with the platforms allowing them to self-regulate: 'Reaching agreements and compromising to fight bad practices is in our DNA' (Catà, 2017). If only Barcelona could be as understanding as Amsterdam (Abad Liñán, 2015)! Finally, Airbnb resorts as usual to its sharing and anti-capitalist rhetoric, although this reproach in the context of left-wing governed Barcelona lacks credibility:

> A darker angle to their frustration is that Barcelona is considered by some to be biased towards the hotel industry, with the renting/sharing model seen as a threat to the status quo and not in the interest of boosting the wider economy and supporting residents who want to earn additional income. (May, 2016)

Remarkably, Amsterdam's policy makers pride themselves on their pioneering agreements with Airbnb and claim that their regulation and control measures are more efficient than in Barcelona, despite the *'tough guys'* rhetoric in the latter city (Jacobs, 2016). This is surprising, not only because the adjustment in the Dutch regulation from a 60- to a 30-day maximum seems to question that efficiency, but especially because Airbnb has used its Amsterdam agreement as a model to be followed in the rest of the world. If we analyse the growth of bookings on the platform as in Table 10.1, the control of urban vacation rentals in Barcelona seems more

Table 10.1 Yearly demand growth Airbnb

	2014–2015ᵃ	2015–2016	2016–2017
Amsterdam	324%	126%	25%
Barcelona	145%	90%	15%

Notes: ᵃComparison November–December 2014 and November–December 2015 (earlier data for 2014 not available).
Source: AirDNA; Bakker et al. (2016a, 2016c, 2017a, 2018b).

convincing, even if we take the local tourism crisis of the final quarter of 2017 into account.

Can a Repressive Policy Be Successful?

Building upon the idea of a municipality seeking conflict rather than agreement, what are the possibilities of controlling urban vacation rentals by enforcing the law? Authorities face two problems when acting against Airbnb providers. With the platform's lack of transparency, there are no instruments to proactively locate listings. The maps on the Airbnb website, for instance, are not precise enough to deduce specific addresses from location data. This means that vacation rentals can only be detected after complaints from neighbours or in massive random searches. Combined with the limited chance of being caught, the financial reward is big enough to compensate for the risk of hefty fines: on average, an Airbnb listing in Barcelona made close to €8000 in 2017, and well over €10,000 in the Eixample neighbourhood. It should be remembered that probably most of these amounts are untaxed, and that 36.5% of Airbnb hosts in Barcelona operated more than one unit in 2017; together, this group accounted for over 70% of all listings.

It is unknown yet how legal actions against the platforms themselves will proceed. The usual reaction of Airbnb is to put the legal responsibility for bad practice or illegal activities on the shoulders of individual hosts, describing its own role as a mere facilitator in the communication process. However, the fact that the platform not only charges commissions to guests, but also business support to hosts – such as insurance and the pricing tool – would be an argument to confirm that platforms are active in creating and offering tourist accommodation. Confronted with legal sanctions, platforms will have to weigh up two conflicting interests: abandoning a destination will be a sensitive blow in the struggle for market share in a 'winner-takes-all' competition; and accepting a restrictive legal interpretation that equates urban vacation rentals to regular commercial accommodation sets a dangerous precedent that may undermine the platform in other cities. We may therefore suppose that platforms will be sensitive to this approach and that it may contribute to cities' forcing them into greater transparency and responsibility.

Urban vacation rental platforms have yet another group of clients: the tourists. Can they be sanctioned if they knowingly stay at an illegal accommodation? Cities would normally hesitate to damage their tourist image by acting against visitors, but this step could become imaginable in a destination that complains about overtourism and that has decided to diminish its appeal for low-cost and party tourists. Since saving costs is the main reason for guests to choose an urban vacation rental instead of traditional accommodation, the threat of being fined could be effective to reduce illegal practices.

A repressive approach would have a number of side-effects. To begin with, it requires a substantial investment in investigation, law enforcement and legal action. Repressive control – door-to-door inspections – may become socially unsettling, especially if we consider that some residents have strong reasons to oppose urban vacation rentals, whereas others have strong financial interests in the business. Besides the direct financial consequences of deterring unwanted visitors, it will greatly affect the visitor-friendly climate. Urban vacation rentals may turn into a hit-and-run operation, in which real-estate owners seek fast and easy money by providing a minimal service to 'money saving' guests.

Fighting the *Easyjetset*: A New Monte Carlo?

Overtourism concerns are directly related to a democratisation of international travel, the growing wealth of the expanding middle classes and more affordable air fares. Is it possible, and is it desirable, to reverse this process? The Catch-22 for cities is that attracting the more affluent tourist modifies the nature of tourist impact, but still leads to commercial gentrification and resident displacement. Our 'exclusivity' scenario described this risk as follows:

> Growing irritation with mass tourism, speculation and the *easyjetset* culminated in 2015, when the city announced a moratorium on hotel investments and a harsh crackdown on illegal short-term rentals. Frozen supply and growing demand allowed existing hotels to raise rates. Since new hotels could not be built, investors concentrated on upscaling 2- and 3-star hotels. The city that until then had been a favourite of nightlife loving millennials and international students, became prohibitively expensive for visitors other than big company expats, wealthy Middle Eastern and Asian tourists and cruise tourists. By 2020, the city had changed into a new Monte Carlo.

> Unfortunately, the new tourists also lifted cost levels in retail. The souvenir stores and fast-food outlets that symbolized the 'old' tourism had been replaced by design and jewellery stores. Despite the positive economic impact this had on the city, the local authorities faced the same challenge even more than before: how to avoid tourism from displacing local residents from the city centre? (Oskam & Boswijk, 2016: 34)

Scenario 2: Outlawing the Platforms and Deterring Low-cost Tourists

We can conclude that this second scenario for the future of urban vacation rentals, in which a destination seeks to minimise its impact by enforcing restrictive regulations, suggests that this approach can be to some extent effective if it targets the various market parties – supply, demand and platforms – at the same time. Strict controls may 'de-touristify' the city by removing illegal low-cost accommodation and by creating a repressive, unwelcome atmosphere. Replacing the party tourists with high spenders may reduce the abuse of residential housing, but not commercial gentrification. Paradoxically, treating platforms as regular accommodation prevents them from converging commercially with the traditional hotel sector: while a small part may be legalised and transformed into regular tourist accommodation, the majority of unregulated apartments will cease to be a viable alternative for mainstream travellers.

Notes

(1) '[…] no estem en contra dels turistes ni del turisme, nosaltres també ho som i sabem que viatjar és una activitat humana que pot ser molt enriquidora. El problema no és el turisme sinó l'actual model turístic. Un model turístic que respon a un model capitalista que està concentrant els beneficis en molt poques mans, que està provocant la destrucció del nostre territori, que està fent augmentar els preus del lloguer fins al punt que ens expulsa dels nostres barris, allunyant a les persones de les seves xarxes familiars i de suport que fa que el treball reproductiu que assumim majoritàriament les dones quedi encara més individualitzat i suposi una càrrega de feina més elevada, que està precaritzant encara més els llocs de feina, que està convertint les ciutats en aparadors i destruint la pròpia identitat dels barris … En definitiva, un model de turisme que genera beneficis per a molt poques persones i empitjora les condicions de vida de la immensa majoria.'

(2) About the impact of Trump's election there are alternative and contradictory facts; it seems safe to say that the country suffered image damage affecting travel, especially from certain countries of origin (Pan, 2018).

(3) Translation of a newspaper heading (El País, 2016).

11 What if the City Regains Control? San Francisco 2025

> Back in the day, we used to test travelers to see if they were worthy of Couchsurfing. I remember meeting a friendly Malaysian in Bulgaria, and shared a train ride with him. Couchsurfing was so new back then, that there were actually only a handful of hosts in Bulgaria, so Noel had never heard of it. But I felt an innate openness, and warmth, within him, so I told him about Couchsurfing. He joined, and quickly became an active user, and later, an Ambassador, in the site's true early spirit. That seemed like natural, organic growth, spread through word of mouth, introduced by people who shared the same ideals. If you were meant to be a Couchsurfer, you would find it. If not, it would remain apart, a subculture in a world of diversity. With time, society would be ready.
> Unfortunately, we live in a society obsessed with growth, and the Couchsurfing management team fell into this same trap. The millionth member joined in 2009. Now, there are five million users, mass media coverage, and even mentions in Lonely Planet. Was it inevitable? Probably. Could it have been done in a way that respected the values that spurred Couchsurfing's initial organic growth. Definitely.
> Coca, 2013

With an estimated total of 10,000 active listings, San Francisco is not an important Airbnb destination in numbers: with a similar population, Amsterdam has three times more. But as Airbnb's home base, the controversies in San Francisco had not only the longest history but also a symbolic value. This explains the interest sparked around the world by the social and political debate around home sharing in San Francisco.

The creation of the new platform did not just happen by accident in San Francisco, as the city combines the different conditions that jointly explain its success: appeal to an innovative public, presence of technological expertise, elevated cost of living and high hotel occupancies. All these factors are illustrated by the anecdotal birth of the platform's predecessor, AirBed & Breakfast: the offer of improvised accommodation during a design conference – with local hotels fully booked – to make ends meet. The explosive growth from this initiative in 2007 caused numerous problems to the city: City Supervisor David Chiu speaks of an estimated number of 100,000 short-term rental incidents per year (Tam, 2014). From an initial prohibition of rentals shorter than 30 days which was hard to enforce, San Francisco adopted a more permissive legislation in 2015, meant to bring the developments back under control in a 'a balanced approach relative to cities like New York that have a full ban on short-term rentals and cities like Malibu that permit the use as of right' (Board of Supervisors, 2014).

The collection of tourist taxes was one of the aspects agreed with cities such as San Jose, Santa Clara, Palo Alto, San Francisco and Oakland between 2014 and 2015. Airbnb and other platforms collected the local fees imposed in each city (14% in San Francisco), thus allowing them to maintain the anonymity of their hosts. In a reversed logic, Airbnb presented this lack of transparency as a service to the city: 'It's good for the government officials who won't have to identify hosts and collect the taxes themselves: we'll do the work for them' (Hantman, 2014). But precisely so as to be able to identify home sharers, for taxation and other law enforcement purposes, San Francisco also required them to register. This compulsory registration was the Achilles' heel of the new rules. Hosts apparently have too much to hide to identify themselves voluntarily: by November 2015, the Office of Short-Term Rentals had received 1082 registration applications, which means that 79.9% of the city's 5378 unique hosts were in non-compliance (Brousseau, 2016). This situation left all other stipulations, such as a 90-day cap or the requirement that hosts were residents of the offered units at least three-quarters of the year, without practical effect.

Proposition F sought to limit the yearly maximum to 75 days, regardless of the presence of the host, to require quarterly rental reports from hosts, and to give tenants unions or residents within 100 feet legal standing to take action against Airbnb listings that violate the rules. A referendum was held on 3 November 2015, in which the Proposition was backed by affordable housing associations and workers' unions such as the California Federation of Teachers and the California Nurses Association. Airbnb invested heavily in a counter-campaign, raising US$8m against the US$486,000 for the campaign backing the measures, and managed to rally a host movement. The Proposition had several weaknesses – an avalanche of civil litigations was feared to be the result, and the measures could only be modified through a new ballot – and was finally rejected by 55% of the votes (Booth & Kiss, 2015). Airbnb's framing of the conflict was twisted as usual: 'Voters stood up for working families' right to share their homes and opposed an extreme, hotel-industry-backed measure' (Said, 2015).

The city then attempted to enforce an obligation to register through the platforms themselves, making them legally liable for offering unregulated listings. This legislation was disputed by Airbnb and HomeAway in federal court, alleging among other things the freedom to publish third-party content. Although the platforms' arguments were dismissed, the judge ordered a stay for the application of the US$1000 fines for each transaction of an unregistered listing. In the following negotiations, Airbnb apparently adopted a more conciliatory attitude, accepting the obligation to require a registration certificate for all listings. It is plausible to suppose that this change of mind led to subsequent concessions by the city, such as a veto of the legislation that would reduce the legal rental cap to 60 days (Cave, 2016; City Attorney of San Francisco, 2016; Donato, 2016).

The dispute was finally settled in April 2017, with the platforms agreeing to delist unregistered units, while simplifying the registration process by incorporating the city's registration system onto their websites. The platforms also committed themselves to disclosing listing information, thus actually enabling the city to enforce the regulations (City Attorney of San Francisco, 2017). At the beginning of 2018 when the changes started to take effect, thousands of listings were reported to have become inactive in the city. According to Host Compliance data, the number of active vacation rentals in San Francisco fell from 10,349 in August 2017 to 4778 in January 2018. The Airbnb share of these rentals was reduced from 8740 to 4191 (Said, 2018a). These snapshots of activity on a certain date of measurement are not comparable, though, to yearly totals, as most listings are not active around the year. The difference between August and January can partly be explained by seasonal differences. Airbnb stated that it had the same number of reservations in San Francisco in December 2017 as the year before, in sharp contrast to the platform's usual double-digit growth numbers (Said, 2018b).

The San Francisco Housing Crisis

In the previous chapters we discussed housing issues associated with gentrification and urban vacation rentals in Amsterdam and Barcelona, but in the case of San Francisco this problem is particularly acute. With a median house sales price of US$1.3m and a median rent of US$4500 (Trulia, 2018), San Francisco is in the same league as New York City, and among the most expensive cities internationally, as is shown in Table 11.1.

However, as with the 'touristification' debate, these high prices cannot be attributed solely to urban vacation rentals, as their evolution goes further back than the birth of the platforms. Supply inelasticity and regulations have been blamed for soaring prices (Glaeser & Gyourko, 2002; Glaeser *et al.*, 2005a, 2005b). Figure 11.1 shows how, corrected for inflation, housing prices rose 162.3% in San Francisco between 1980 and 2016, beating New York. In both New York and San Francisco, house prices have risen sharply since the 1990s, with the exception of the crisis years (Figure 11.2).

Table 11.1 Rents and property prices (2018)

City centre of:	Rent 1 bedroom apartment	Rent 3-bedroom apartment	Price/m² to buy
Barcelona	$1069	$1737	$5,498
Paris	$1320	$2853	$11,786
Amsterdam	$1800	$3060	$7,559
Los Angeles, CA	$2098	$3680	$6,631
Washington, DC	$2108	$3954	$6,036
Singapore	$2151	$3812	$18,403
London	$2202	$4150	$17,806
New York City	$3138	$6171	$13,490
San Francisco, CA	$3305	$5492	$11,773

Source: Numbeo (https://www.numbeo.com/property-investment/).

Finally, we can observe in Figure 11.3 that soaring house prices have long outperformed annual rents, but that since the beginning of 2014 these amounts have started to converge for San Francisco, while New York rents do not rise as fast as house prices.

In Figure 11.4, McCann (2015) compares the evolution of two-bedroom apartment rents in the city with the dot-com bubble and finds the steepest increase after 2005. The graph in Figure 11.4 suggests that the gentrification processes in San Francisco are closely associated with the

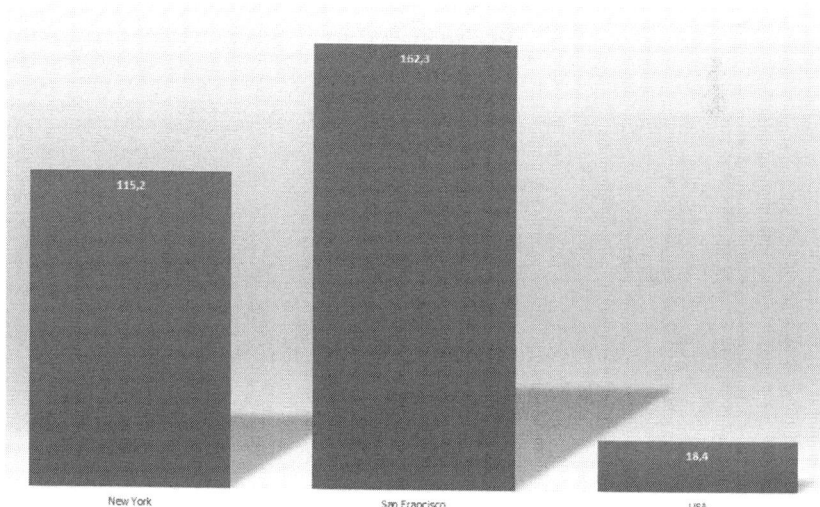

Figure 11.1 House price percentage change, in real terms
Source: Zillow; Bureau of Labour Statistics; The Economist (2016).

Figure 11.2 House price index, Q1 1980 = 100
Source: Zillow (The Economist, 2016).

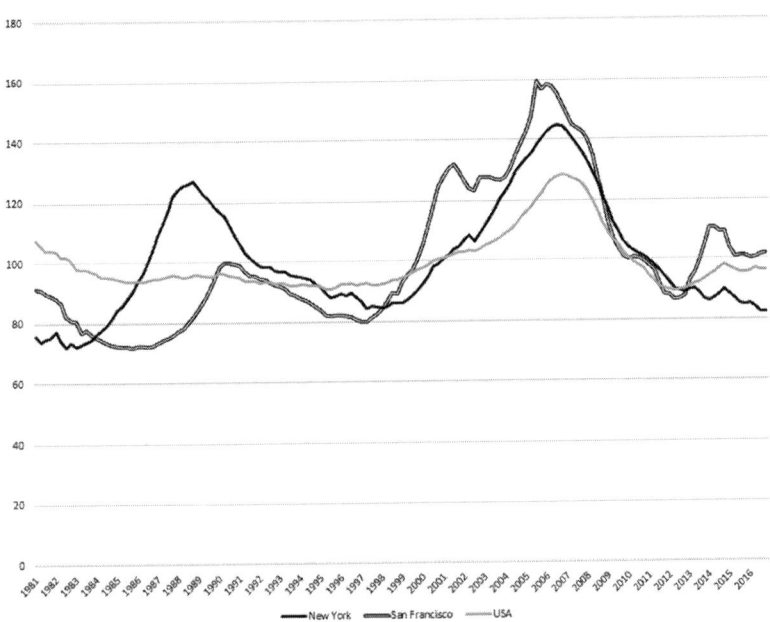

Figure 11.3 Ratio of house prices to annual rent, long-term average = 100
Source: Zillow; The Economist (The Economist, 2016).

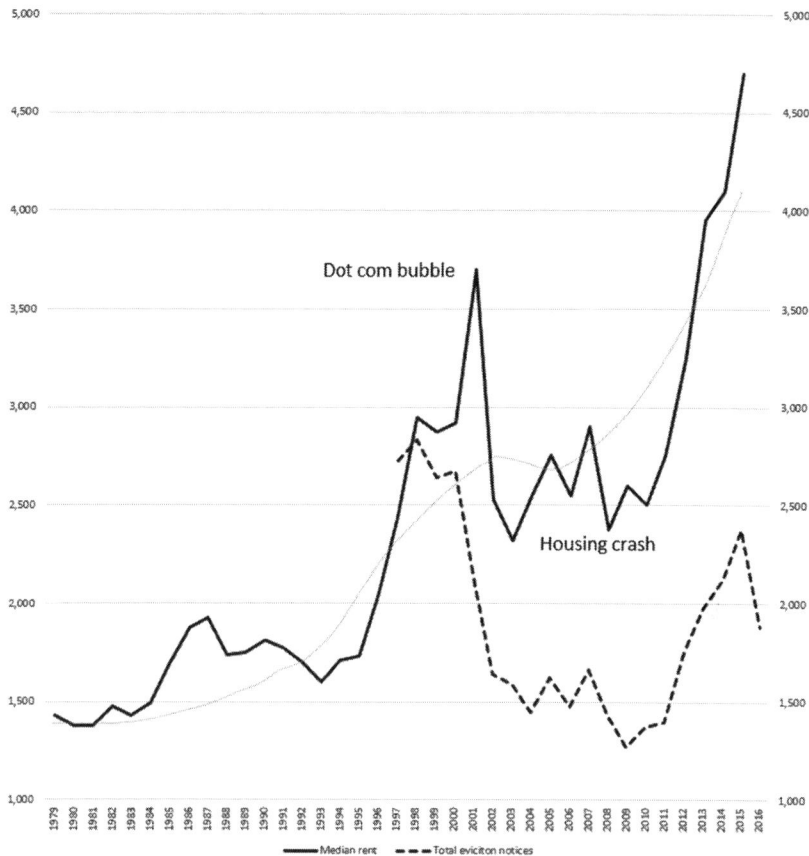

Figure 11.4 Median monthly rent price of a two-bedroom rental in San Francisco, corrected for inflation[1]; and total eviction notices 1997–2017[2]
Source: [1]McCann; [2]San Francisco Residential Rent Stabilization and Arbitration Board.

rise of the tech industry in Silicon Valley and in the city itself. The average salary for a Silicon Valley tech worker is US$101,278 as opposed to the general median income for one person in San Francisco of US$70,850 (Goldman, 2013). The displacement of low-income residents – in relative terms, not compared to US national averages – is illustrated by certain trends in the housing market. First is the rise of housing types that would previously have been considered below standard. The minimum living space was reduced to allow the construction of 220 square feet (20 m²) micro apartments, for rents that were expected to range from US$1000 to US$1500 (Hurst, 2012). An alternative is 'dorm living', where the disprivileged with incomes ranging from US$40,000 up to US$90,000 can rent 130–220 square feet (12–20 m²) bedrooms with shared bathrooms for US$1400–2400 per month (Bowles, 2018). Secondly, a direct indication of displacement is the number of evictions, which also according to Figure

11.4 showed a steep increase from 2011 to 2016 (SF Residential Rent Stabilization and Arbitration Board, 2017).

The seriousness of the housing problem in San Francisco helps to explain the strong presence of a counter-movement in tenants' unions and affordable housing associations, which would also become important in the opposition against vacation rentals. A famous episode in the unrest and opposition produced by gentrification was the so-called 'Google Bus Protests', which occurred from the end of 2013 to 2016. The 'Google Bus' had become a generic name for the free shuttle buses offered by Google or any other tech company to its employees to facilitate an inverse suburbanisation: the new generation of tech workers would live in the city centre of San Francisco and commute to the Silicon Valley outskirts for work. They had therefore become symbolic of the impact of tech-driven gentrification. But the Google Buses also had a real effect on gentrification: proximity to Google Bus stops is commonly listed under the amenities of rental apartments in San Francisco, and Goldman (2013) demonstrates that in most cases apartments within walking distance of Google Bus stops have increased rents more than elsewhere.

San Francisco Tourism Statistics

In 2017 San Francisco was estimated to have received 10.3 million overnight visitors and 15.2 million day visitors, up from, respectively, 9.6 and 13.3 million in 2013. The city has approximately 218 hotels with 34,002 rooms. Hotel occupancy and average daily rates had decreased slightly compared to 2016, from 85.4% to 83.3% and from US$252.92 to US$249.49. As everywhere, visitor statistics are mainly based on hotel guest data; users of urban vacation rentals are probably not fully accounted for (Allor, 2018a, 2018b).

Airbnb protests in San Francisco revolve around housing and gentrification issues, not around excessive tourist pressure. Curiously, rather than tourists displacing locals as in other cities, in San Francisco the complaint is that poor people drive away the visitors: the growing homeless population – 7499 in 2017, up from 6436 in 2013 – was blamed, along with the declining image of the United States under the Trump administration and the works at the city's convention centre – as one of the causes of disappointing tourist numbers (Brinklow, 2018; Littman, 2018; Spotswood, 2017).

Genuine Sharing

The utopian narrative used to promote urban vacation rentals almost makes us forget that Airbnb and similar platforms are not the inventors of 'home sharing', just of turning it into a profitable business model. One of Airbnb's predecessors was Couchsurfing, a platform to connect travellers looking for a place to stay on their trips. Couchsurfing transactions can be considered as actual 'sharing', both in the sense that they intensified the use of an underutilised asset – space to live, or a couch – and that they were meant to establish a genuine contact between guest and host. One of the conditions of the platform was that transactions were non-monetary, so that hosts were not allowed to charge for offering accommodation. The platform itself depended on donations and on a so-called 'freemium' model in which a basic service is offered for free and users pay for advanced features: in the case of Couchsurfing, users could opt for a verified membership at a cost of US$25.

In 2011, Couchsurfing decided to give up its non-profit status, a decision that was justified by the bureaucratic hurdles to innovation derived from its reporting obligations. The platform was converted into a for-profit business and raised US$7.6m in funding (Perloth, 2011). Even though the company's management declared that there was no intention to change the terms of service, the changes and suspicions of commercialisation sparked outrage among the user community. Advertisements were placed on the site as a test to create a revenue stream (Billock, 2015). The size of its user base has almost tripled since 2013 to its current 14 million members (Couchsurfing, 2017). One of the often voiced complaints is that the community spirit had been sacrificed to growth: the website had incorporated other social media functionalities, and new members were accused of using the site as a dating tool rather than for its original purposes (Coca, 2013; Johnson, 2011; Lewis, 2013).

An alternative for alienated couch surfers was the French site and organisation BeWelcome, which fully depends on donations and volunteer work. The site offers a free service for non-monetary home sharing. In 2018 it had slightly more than 100,000 users, mainly in European countries (13% German, 11% French and 10% United States) (BeWelcome, 2018).

An idealistic home-sharing initiative, founded after WWII with the mission of 'fostering understanding of cultural diversity through a global person-to-person network promoting a more just and peaceful world' is Servas, Esperanto for 'we serve' (meant as: serving peace). This non-profit platform aims to bring people together for the sake of pacifism, and charges a membership fee to cover its overhead costs. The organisation has 15,000 member hosts and 'thousands more travelers' (US Servas, 2018).

Guest to Guest was founded in France in 2011 as a home-exchange platform, with the peculiarity that home swaps can either be reciprocal or indirect, through a points system: credits received for hosting can be used to 'pay' for accommodation at a different time. Through a deposit system, the platform aims to enhance the security and trust of its hosts and guests; Guest to Guest itself is funded out of the 3.5% commission charged on these deposits. Apart from the safety deposit, the home exchange is a non-financial transaction. The platform has 300,000 members in 187 countries, mostly concentrated in France, Italy and Spain (Guest to Guest, 2018).

The San Francisco-based BeeNest has introduced a blockchain-based decentralisation as a solution to eliminate the 'middle man' of platforms like Airbnb and to offer decentralised home sharing. High commission fees and security breaches caused by platform growth are the main disadvantages of mainstream platforms addressed by this blockchain alternative. The platform is still in alpha stage and does not report membership numbers (Bee Token, 2017).

These sharing alternatives seek to reinvigorate the ideals of empowering residents and offering a 'living like a local' experience while eliminating the possibilities for commercial operators as well as a profitable intermediary platform. Except for Couchsurfing, they also share the characteristic of being marginal in membership numbers compared to the incumbent 'home-sharing' platforms. Is a turn of events imaginable which would stall the 'winner-takes-all' platform competition dynamics and bring 'genuine sharing' initiatives to the forefront?

A Grassroots Counter-movement?

If we think back to the teachings of Airbnb executive and advocate Douglas Atkin, a cult community can only exist because of its differences from what it considers as outsiders. The quote from a disappointed couch surfer that opened this chapter shows how community growth can erode the community feeling of early adopters: when the underground movement even gets mentions in Lonely Planet, it can no longer be authentic. It is conceivable that some of the early adopters of Airbnb share this same feeling of alienation. But unlike the original couch surfers, they have strong incentives to stay loyal to Airbnb: their reputational capital is stored on the platform, they may have built up a regular income that supports their living expenses, and no alternative can offer the same client base to guarantee the continuation of that income.

In other words, the resurgence of genuine sharing can become possible if an event occurs that seriously dents the oligopolistic market power of Airbnb. Such an event could be: (1) the loss of its platform function because of decentralised communication between host and guest; (2) a substantial limitation of one of its customer bases that obliges the other side to look for alternative suppliers; or (3) the exclusion of clients, forcing them to look for an alternative. The emergence of blockchain technology promises that the first condition will occur, although it is still hard to predict its real impact. Conditions 2 and 3 may have already been created, as closely interrelated circumstances, by the recent crackdown on illegal or unregistered Airbnb rentals.

Transparency and registration are deterrents for hosts who engage in illegal activities such as tax evasion or unapproved subletting, for hosts who make a limited use of the platform, or for hosts that offer accommodation without achieving any bookings, for instance because their apartment is in an unattractive neighbourhood. The first category does not need much explanation: forced to open their books to local authorities, these hosts will also be detected by national tax services or by their landlords. All other hosts will weigh the costs and benefits of registering their unit. Occasional earnings may become insignificant when they become taxed. Besides decreased financial efficiency, the bureaucratic hurdles will widen the gap between professional and occasional hosts.

We can calculate the approximate impact of the regulations on a host's rental income. The average daily rate (ADR) in San Francisco in January 2018 for all types of properties was US$243 (Said, 2018a). Airbnb commissions are 3% of the revenues. Airbnb will also retain the San Francisco Transient Occupancy Tax of 14%. Other federal and state taxes will vary depending on the situation of each host; for this estimate we will use an average of 29.8% (Frankel, 2017). The registration process requires hosts to purchase a US$91 Business Registration Certificate, which may be reimbursed if only used for short-term rentals. The Short-Term Residential Rental Certificate registration fee is US$250 for two years; this fee is non-refundable, even if a host's application is denied (Airbnb, 2017). As shown in Figure 11.5, this would mean that a compliant occasional host who rents out his apartment for 10 days sees that his average US$2430 revenues are reduced by 56%, leaving the host with US$1075. If an apartment is rented out successfully for 30 days, the host's share is still reduced by more than 46%, from US$7290 to US$3907. ADRs may be higher than this January average if the host chooses a rental period strategically; however,

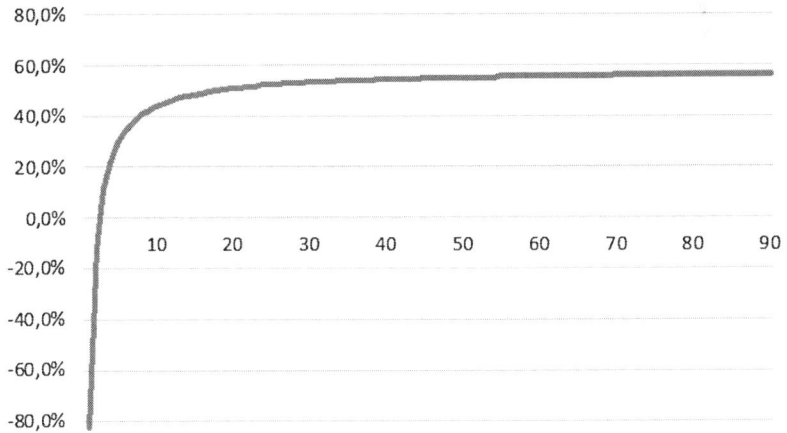

Figure 11.5 Estimate of net rental revenues per night for regulated San Francisco short-term rentals

research shows that occasional hosts achieve lower revenues than professionals (Li *et al.*, 2015; Oskam *et al.*, 2018). The conclusion must be that economies of scale will result in higher concentration and professionalisation, under these circumstances.

For the occasional host, genuine sharing may become more appealing under certain circumstances. If we suppose that additional vacation rental income is spent during the host's own holiday, the revenues of a full 30-day rental activity will be sufficient to stay 26 days in Amsterdam with an ADR of US$166, or even 34 days in a cheaper destination such as Barcelona with an ADR of US$112. But this is only because of the price difference between cities, strongly in favour of San Francisco; the more this price gap is closed, the more appealing it will become for occasional hosts to barter accommodation days.

The financial discouragement must be added to the immaterial advantages of genuine sharing, especially if the hosts pertain to the user segments of 'collaborative consumers' or 'interactive novelty seekers' (Guttentag, 2016). The result could be that, while professional hosts stick to Airbnb or other commercial platforms, the non-professionals start moving to non-monetary home-swap alternatives. Innovations, such as the flexibility created by Guest to Guest's points system or Beehost's decentralised transactions, may contribute to such a non-monetary platform getting a similar, or even a more productive functionality than the commercial providers.

> ### Scenario 3: Commercial, Hotel-like Rentals Versus a Revival of Genuine 'Sharing'
>
> In this third scenario, we will see a full convergence between hotel accommodation and regulated urban vacation rentals, due to a concentration and a further professionalisation driven by advantages of scale. Non-professional hosts will be driven off this market increasing the popularity of non-monetary, genuine sharing home-swaps. The regulation of vacation rentals will not reverse the city's housing shortage, as it is not its only cause, but it will give the city further instruments to control the housing market and to develop the policies its residents choose.

12 What if the City Outsmarts the Market? Singapore 2025

> When dental assistant Rodelita C. Leng's family members travel, they use Airbnb. Yet, the 38-year-old resident of Vacanza@East in Kembangan would not like the same to happen here. 'If my neighbours register their units under AirBnb, anyone can just go in and out. This is not safe and secure for my two young children,' she said.
>
> Heng and Jo (2016)

Singapore is Airbnb's bridgehead to conquer the Asian market. In 2013 the company opened its regional headquarters in the city state, because of the availability of a skilled workforce and the city's good connections and infrastructure. After one year, it employed 30 people from all over Asia. Singaporeans were also the platform's main Asian market to book accommodation elsewhere. Rather inconveniently, hosting Airbnb units had not really taken off: on the contrary, renting out rooms or apartments for less than six months was punishable by fines of up to US$160,000 or even a 12-month prison sentence. But as a business publication announced, 'change may already be underway' (Hartung, 2014). This was in 2014; four years later it is safe to assume that it was not.

Two explicit concerns motivate Singapore's reluctance to allow urban vacation rentals. In the first place is the threat holiday-goers could pose to the living environment and safety of residential neighbourhoods, either because of their higher rate of turnover or because of minor offences to the orderly life of the estates. As one resident declares: 'Some Airbnb users throw their cigarette butts and cause a lot of inconvenience' (Heng & Jo, 2016). A second concern is that the service inconsistency of urban vacation rentals might negatively affect the image of Singapore as a safe, clean and well-organised destination.

Besides its reputation as an economically successful, crisp and business-friendly oasis, Singapore has invested heavily in IT education and infrastructure, and has achieved worldwide recognition for its development as a 'smart city'. Urban transport and waste collection managed with big data, self-driving buses, easy online access to government services, an alert system for first aiders in cases of out-of-hospital cardiac arrest, sensor networks to measure and regulate temperature and humidity, and elderly monitoring systems are among the projects and initiatives envisioned by Singapore Smart Nation (2018).

Tourism in Singapore

Singapore is a prominent example of the Asian 'Tiger' economies, with an average yearly GDP growth of 6.8% between 1975 and 2018, closely managed in a paternalistic political system controlled by the hegemonic People's Action Party. As a business magazine euphemistically comments: 'Because Singapore never developed a culture of political rights under British imperialism, the newly elected PAP sought to legitimize itself not by protecting rights – the basis for the state in western thought – but rather by overseeing robust economic growth' (Tim & Sussman, 2014). Subsequently, government actions have aimed at fostering trade and attracting international businesses, developing an educated workforce – especially in the field of IT – and promoting an entrepreneurial attitude. Singapore is currently second after New Zealand on the World Bank's 'Ease of Doing Business' ranking, a list it headed for 11 consecutive years. Between 2011 and 2014, the country was second only to Taiwan in its residents' intent to start a business within the next three years. Its GDP per capita ranked 14th globally with US$52,962, between Denmark and Sweden (Chernyshenko et al., 2015; Trading Economics, 2018; World Bank Group, 2017, 2018).

The city state has a resident population of almost 4 million, plus an additional non-resident population of 1.65 million. In 2016, it had 16.4 million international visitor arrivals, which gives a considerably lower tourist pressure – expressed in a visitor per inhabitant ratio – than the other cities we have examined. In view of the reputation of the island as a business hub, it may be surprising that leisure visitors by far outnumber business and MICE travellers, with 69% to 15%. Almost half of the visitors stay at or go to Marina Bay Sands or Resort World Sentosa, so-called integrated resorts including hotels, convention facilities, entertainment shows, theme parks, luxury retail and fine dining, which constitute a strategically positioned offer to enhance Singapore's destination appeal. The largest visitor markets are Indonesia (17.6%), China (17.5%) and Malaysia (7.0%), of which China had the largest growth: 36% in 2016, and an average of 11.1% in the last 10 years.[1]

The cost of living is higher in Singapore than in the surrounding countries, making it a relatively expensive destination (Table 12.1). Expenditure data confirm this image of an upmarket shopping destination. Visitors' largest budget items of their US$1145 expenditure per capita are shopping (23.1%) and accommodation (23.0%). The most popular shopping items are, by far, clothing and accessories (41% of amount spent, 58% of

Table 12.1 Big Mac Index and hotel average daily rates (US$) for Singapore, Malaysia, Indonesia and China

	Big Mac index	Hotel average daily rates
Kuala Lumpur	2.28	135
Hong Kong	2.62	222
Bali	2.68	146
Jakarta		166
Sanya	3.17	386
Shangai		153
Beijing		142
Macau		169
Singapore	4.39	187

Source: JLL (2017); The Economist (2018).

Table 12.2 Percentage of total and of shopping expenditure per country 2016

	All	Indonesia	Malaysia	China
Shopping	29	34	33	43
Shopping amount spent on fashion and accessories	41	48	52	40
Shopping amount spent on cosmetics, perfumes, etc.	17	17	15	23
Accommodation	29	18	23	25
Food and beverage	14	11	14	9
Other	28	37	30	23

Source: Singapore Tourism Board (2018).

popularity). This is especially true for visitors from the surrounding countries, as shown in Table 12.2.

With an average length of stay of 3.4 days, these tourists total 56 million visitor days. Singapore had 236 'gazetted' hotels – in compliance with the standards of the Singapore Tourism Board – and 58,876 rooms by December 2016, an increase of 34% compared to 2012. Hotel performance peaked in 2012 and has slightly decreased since then, in terms of occupancy (from 85.7% to 83.1%) and especially of revenue per available room (RevPAR, from US$155 to US$137). Economy hotels are most strongly affected, as shown in Table 12.3.

The majority of overnight stays were in hotels, although less so in the case of visitors from Malaysia and Indonesia (Table 12.4). Regional Airbnb Director Chai Jia Jih referred to the Asian habit of staying with friends or family on their trips as an opportunity for urban vacation rentals (Hartung, 2014). Whether that assumption is true is difficult to tell, but it contradicts one of the explanations of the failure of Airbnb on mainland China (see Chapter 2).

Table 12.3 Hotel performance in Singapore 2016

Hotel tier	Standard average occupancy rate	RevPAR (US$)	Change
Overall	83.1%	137	– 4.9%
Luxury	85.1%	282	– 0.3%
Upscale	84.7%	164	– 2.7%
Mid-tier	84.6%	106	– 3.7%
Economy	77.2%	58	– 4.8%

Source: Singapore Tourism Board (2018).

Table 12.4 Type of accommodation in Singapore 2016 per country of origin

	Total	Indonesia	Malaysia	China
Hotel	61%	47%	45%	67%
Friends or relatives	18%	25%	31%	17%
Others	21%	28%	24%	16%

Source: Singapore Tourism Board (2018).

Airbnb Regulations

Even if there might be a cultural acceptance of staying at an urban vacation rental, as the evolution of Chinese platforms suggests (as well as the opening quote of this chapter), people may be less open to the idea of letting strangers use their own residence. Furthermore, short-term rentals have been illegalised in Singapore, as mentioned earlier, mainly to protect the living environment of residents and the peace and safety of their neighbourhoods. In 2015 the Urban Redevelopment Authority (URA) opened up the possibility of allowing certain types of short-term rentals to public consultation, which after a year was extended. In the meantime, the prohibition was maintained, and recently the pressure on Airbnb hosts has increased.

In line with the reasons to prohibit short-term rentals, the first reports of actions taken against Airbnb hosts were triggered by complaints from neighbours or community managers. From 2014 to 2016, a total of 1000 properties were investigated; in 2017 the pace was stepped up to 600. Enforcement action was taken against over 100 owners between 2014 and 2016. Despite the hefty penalties, the enforcement action implied simply receiving a URA notice: as an Airbnb host posted on a forum, he 'was invited to tea at URA, except none was served'. Most hosts were reported to have desisted from offering their property after this intervention (Mahmud, 2017).

In 2017 the minimum rental period was cut from six to three months; the new rules also dictated that up to six unrelated people were allowed to stay in a private residential property at any one time (Channel News Asia, 2017; Singapore Legal Advice, 2017). A more recent URA proposal suggested authorising short-term rentals provided that owners representing at least 80% of the development's share value agreed (Wen-Yi & Wei, 2018).

The enforcement actions and these regulation adjustments reveal that an active offer of Airbnb listings has evolved in Singapore: according to the platform's data, an increase from 7781 in May to 8601 in December 2017, hosting 350,000 guests. During the September Formula 1 races, Airbnb use surged 40% (Hong, 2018; Jun Sen, 2017; Liang, 2018). AirDNA detects around 7500 active units – simultaneously active, meaning that these numbers are lower than yearly totals – offered by 2555 hosts, 792 of whom are multilisters (AirDNA, 2018). How can they operate? The first explanation is simple: as long as there are no complaints, no action is usually taken, resulting in a tougher screening of guests by Airbnb hosts in Singapore (Mahmud, 2017). 'Deception and evasion' might be required, such as 'leading guests to a different unit to wait out suspicious security guards who had followed them, or asking guests to lie about the actual unit they were staying in' (Au-Yong, 2018). A legal workaround, although not recognised by the URA, is the 'diplomatic clause' which allows for an early termination of apartment leases without penalty (Khoo, 2016).

The end of 2017 saw the first court case against Airbnb hosts. Two real-estate agents were convicted with fines of US$45,800 each and had their licences revoked for renting out four units through the platform. They were found to have set up a number of companies that rented four units in a condominium, which were then sublet to tourists for short-term stays. The charges show that they pocketed US$3285 for renting a 59 square metre flat to two guests for 23 nights; six guests paid US$775 for five nights in a flat of the same size. The prosecutor argued that, as these amounts corresponded to the period between 15 May and 21 June 2017, their actual profits must have been much higher since the operation had started in late 2015 (Associated Press, 2018; Au-Yong, 2018).

The discussion on regulation is still ongoing. The same principles are maintained in the current public consultation: exclusion of public (Housing Development Board) housing, a 90-day rental cap, a six-person occupancy cap and consent of 80% of the owners. Besides, new proposed rules foresee an obligation to register, an obligation for hosts to keep guest records and a greater liability for the platforms themselves for the legality of advertised listings (URA, 2018).

Smartness

In 2017, Singapore beat cities such as New York and London and topped the Global Smart Cities Ranking, a project sponsored by Intel. The report, titled *Smart Cities – What's in it for Citizens?*, highlights the advantages for citizens of technology-driven solutions in areas such as mobility, health, safety and productivity (Juniper Research, 2018). But the sponsorship raises a second question: What's in it for Intel?

A city's classification as 'smart' does not, in fact, respond to any academic criteria or definitions, and the designation has multiple meanings or interpretations. This definitional impreciseness has allowed it to become in many cases a 'self-congratulatory' adjective, while at the same time it conceals an ideological perspective: the idea that urban problems are resolvable by technology or, in other words, that cities are not rocked by conflicting social interests but just by a lack of technological progress or investments (Hollands, 2008, 2015).

The debate on the 'smart city' concept is centred around a 'hard', technology-driven versus a 'soft' approach, with greater attention to the societal needs and implications of innovation. In general, Asian cities share the political conditions and also the problems that favour a 'hard' smart policy: a centralised government supportive of private investments, and traffic congestion and pollution as urgent issues to address. Singapore is therefore a perfect example of the smart city that 'senses and acts', that collects data to organise and streamline the efficient use of resources (Angelidou, 2014; Nam & Pardo, 2011; Neirotti *et al.*, 2014; Zubizarreta *et al.*, 2015).

Singapore's Infocomm Action Plans – multiple year plans for the implementation of 'smart' policies – have focused in the last decade on citizen or 'customer' engagement: a corporate co-creation strategy, in line with the country's paternalistic politics, to build 'an interactive environment where the Government, the private sector and the people work together seamlessly' (IDA, 2011). The URA's public consultations about short-term rentals were, in fact, examples of this collaborative government. Communication with government services is, however, part of a broader intervention in public life meant to increase efficiency, safety and competitiveness: smart, connected traffic solutions have reduced congestion and the number of cars in general; remote healthcare for elderly citizens ensures that the healthcare system does not become overburdened; smart video surveillance has been trialled to detect criminal activity; and innovation in the private sector is encouraged through funding as well as specialised test-bed environments (Juniper Research, 2018).

Controlled 'Sharing'

We can conclude that Singapore's governmental agencies have studied the disruptive innovation brought by urban vacation rentals with keen interest, but after careful consideration have decided that the advantages are outweighed by the risks to the order, safety and cleanliness of its citizens' neighbourhoods. How will this decision influence the future evolution of the short-term rental market, in view of the characteristics of tourism in the city and its 'smart' policies?

As we have seen, the will to regulate is usually hampered by the enforceability of the rules, with lack of access to host and platform data being the critical factor that limits authorities to reactive enforcement, such as after complaints from neighbours. A city that 'senses and acts' will have multiple ways to break through platforms' secrecy, even without forcing them to more openness. There is a range of surveillance and control devices or systems that have been trialled, ready to be implemented, particularly there where invasions of privacy by authorities are not met with citizen protests: examples are video surveillance with facial recognition; border security with automated lie-detecting kiosks; and different types of behavioural analysis (Daniels, 2018; Denyer, 2018; Lee, 2013; Poon, 2017). A further development of citizen control is no longer the domain of paranoid science fiction: telephone tracking, credit card spending or water and energy consumption patterns will allow the prediction of whether a tenant is a resident or a tourist. Insights on Airbnb multilister investments can be matched with housing market transactions. And why could we not analyse waste water for DNA traces to establish how many unrelated individuals have used a single housing unit?

However, if we consider that the Singaporean authorities are concerned about the disturbance of the peace and quiet of their residents but not averse to private entrepreneurship, an alternative scenario could also be imaginable: one in which smart techniques are used as a matchmaking role, selecting the right tourist for the right environment. This would imply that the guest screening, currently done as a precaution by illegal hosts, becomes automated and centralised as a condition for more permissive regulations. Behavioural control and guidance of tourists could be incorporated into the systems to prevent undesirable side-effects of their presence in residential areas.

Scenario 4: Platform Hospitality Under Surveillance

Both these 'smart' scenarios – strict prohibition or controlled admission – are driven by a destination's capacity to fully monitor and control its citizens' and visitors' behaviour. With the obstacle of platforms' lack of transparency out of the way, the city becomes able to manage visitor streams with a level of detail unseen so far. While this scenario may seem more plausible in a controlled regime such as Singapore than in more open societies, data collection and usage in a city that 'senses and acts' will always bring about these risks or this potential. The goal may not necessarily be to prevent the externalities of urban vacation rentals; the 'entrepreneurial city' may also seek to optimise the benefits of available housing. In the case of Singapore, the importance of other economic sectors and the limited availability of alternative space are good reasons to preserve and safeguard residential areas by banning short-term rentals. Under different circumstances, a smart city may decide for a full convergence of urban vacation rentals and traditional accommodation by creating and managing an integrated offer for its visitors.

Note

(1) The data (Singapore Tourism Board, 2018) do not include Formula 1 visitors.

13 Conclusion: A Neoliberal Nightmare

These four scenarios are meant as a contribution to the social debate around urban vacation rentals. The many questions that have arisen after its virulent invasion into city centres – significantly more than anywhere else – can be summarised as: is it a good or a bad thing? Of course it would be extremely controversial to simplify the analysis of a business innovation to a one-dimensional ethical judgement. However, in the case of Airbnb – the genericised trademark for urban vacation rentals – this has become unavoidable: the brand has been marketed as something 'right'. Its moral standards, derived from the ill-defined concept of 'sharing', were supposedly higher than other commercial propositions: it was anti-consumerist, it re-enabled travellers to find authentic human contact, it saved the environment and it helped poor people.

The rhetoric of 'sharing' has also identified the evil it is fighting: big business. Abundant examples in Airbnb statements oppose their 'right' to the 'wrong' of the hotel industry. We should be extremely cautious about adopting that false distinction. While there is certainly a conflict of interest between the newcomer and its incumbent competitors, this is nothing unusual in business. But as an economic activity, Airbnb and similar platforms are not opposed to, but part of the hotel industry; their growth has implied an expansion of this industry into new areas and aspects of our lives.

The pessimism of the scenarios leaves little doubt that I expect the impact of urban vacation rentals to be harmful. The practical implication of this is obviously that the phenomenon is a potential 'evil' that cities should not promote or embrace, but regulate – as they have regulated and should keep regulating the externalities of other parts of the hotel industry. But equally interesting is the academic interpretation. How should the 'sharing' narrative and the explosive growth of urban vacation rentals be seen in the light of larger, socio-economic and ideological, developments?

In general, despite a limited number of valuable publications, the academic contribution to this debate has been disappointing. Too many studies have focused exclusively on the business analytics of urban vacation rentals. Even though they often allude to 'sharing' theories, it is not rare for these articles to conclude with business recommendations for vacation rental hosts. At least, they could have been expected to notice the incongruence between commercial exploitation and some definitions of 'sharing'.

Debunking the Platforms' Narrative

Platforms are not communities. They are matchmaking businesses that make a profit by facilitating communication and transactions between vacation rental operators and travellers. In the case of Airbnb, a strong marketing narrative appealing to an 'off the beaten track' lifestyle and travel style has induced a community feeling among its users, not dissimilar to the community spirit that unites believers of 'cult brands' such as Harley Davidson or Apple. It produces fanatical brand loyalty.

Airbnb has an advantage over competing platforms that have not created a loyal follower cult, as both client groups – guests and hosts – are in a position to 'multi-home' or conveniently pick a platform for each occasion and for each transaction. This generates a centripetal force attracting users to the market leader, for the biggest marketplace has the strongest matchmaking power. Unless an external event changes the market-leader appeal, this effect leads to a 'winner-takes-all' competition.

The cult marketing has obfuscated our understanding of user motivations and segments. Airbnb travellers are a heterogeneous group. Although a segment of the users – approximately a fifth – are actual believers in collaborative consumption, the majority are driven by the tangible and intangible advantages of apartment rentals over traditional accommodation. These advantages may be lower price, homely feel, square metres or apartment amenities, or the novelty of staying in a different place or of meeting new people.

Nevertheless, meeting locals – other than the tourism service providers – or experiencing the genuine 'back' instead of the tourist 'front' of local culture, is neither the result nor the objective of many Airbnb trips, for the promise of 'living like a local' refers to travellers' existential authenticity and not to the authenticity of the visited place. Airbnb guests can be their true selves and relax with their travel companions, without the pressure of obligated visits and activities associated with mass tourism.

The image of urban vacation rental providers or 'hosts' is surrounded by the most misunderstandings or deliberate mystifications. Airbnb has claimed to empower poor people in peripheral neighbourhoods so that all, and not just the hotel industry, can benefit from the tourist economy. This emancipatory claim is counter-intuitive. When it comes to renting out people's own residences, one would expect that privileged people can offer more attractive facilities and locations. An entirely free and uncontrolled market will not only sharpen socio-economic divides; it has also been shown to create the conditions for racial discrimination to emerge.

While it may be true, as Airbnb argues, that in most cities around 90% of hosts offer only one unit on the platform – presumably, their usual residence – that number is totally irrelevant. The concentration of Airbnb properties is hidden in the remaining 10% minority of hosts who, depending on local circumstances, represent one-third to two-thirds of the Airbnb offer in their city, and also consistently an even larger share of bookings and revenues. These 'multilisters' are an indication of the professionalisation of hosts, either because investors enter the market with professional expertise, or because successful Airbnb hosts quit their day jobs and professionalise. The evolution of multilisters in London is an indication of how this accumulation process works.

As for the claim that urban vacation rentals, and specifically Airbnb, contribute to spreading tourists to peripheral areas, that has been proven simply and consistently untrue. Tourists prefer central areas and, unhindered by zoning plans or other regulations, the demand becomes strongly clustered in the city centres and around tourist landmarks. Even when an offer appears in less attractive and peripheral areas, as the less privileged attempt to participate in the benefits of tourism, the demand for these accommodations is limited or zero.

Lately, Airbnb has revealed strategic moves by inviting boutique hotels and bed & breakfasts to join the platform, in addition to vacation rentals offered by named – and ostensibly non-professional – hosts. This strategy eliminates the cultural or marketed gap between traditional hotels and vacation rental platforms. Airbnb's convergence also implies incentives to hosts to standardise and professionalise their offer, thus further eroding the image of vacation rentals as something opposed to mainstream commercial hospitality. The abandonment of the divergence narrative will remain controversial to the most faithful believers in 'collaborative consumption', and probably indicates a calculated risk of image loss for the sake of expansion beyond its current markets.

A Grim Future: Why is There No Positive Scenario?

The future of the platforms is determined by their strategies and by the evolution of consumer behaviour, but also by the commercial counter-strategies of traditional hotel companies and by authorities seeking to control the phenomenon. This control is aimed at protecting the authorities' self-interest, by acting against tax evasion, but also the interest of traditional hotels which have demanded a 'level playing field', and the interests of residents to counter the external effects of vacation rentals, i.e. noise and other nuisance, rent hikes and overtourism. So far, regulations have seldom been effective as authorities lack sufficient information to enforce the rules.

Inspired by Dator's 'alternative futures framework', four scenarios were initially derived from the standardised outcomes of the *continued growth* of a trend or development, its *collapse*, its acceptance but without its excesses by a *disciplined society* and its incorporation into a *transformational society*. This led to examples of: Amsterdam overrun by mass tourism; Barcelona cracking down on illegal rentals; and a disciplined society, hypothetically located in San Francisco, that would return to the idealistic model of non-monetary sharing, with Airbnb evolving, in full convergence with the rest of the hotel industry, as a regulated accommodation provider. The transformational society that successfully addresses the issues that arise because of the new developments through technological solutions – in the example of Singapore – can opt for eliminating the phenomenon or for having it converge with its existing business structures.

Actually, it turns out that these four imagined futures could be placed in a conventional scenario cross, with two uncertain factors leading to different outcomes. In the first place is whether a city decides to embrace or to some extent tolerate urban vacation rentals, or seeks to eradicate them as illegal. In the second place is whether platforms can be deprived of their secret weapon: their lack of openness. Once cities succeed in forcing platforms such as Airbnb to greater transparency, they put an end to their exceptional, unregulatable position as 'non-businesses' that exist in a parallel universe without zoning plans, tax inspections or consumer protection.

Nevertheless, all four scenarios are rather pessimistic. Is there no other option than losing our cities to tourists or hunting them down as criminals in a police state? Can altruistic sharing only revive after immiseration has left it as the only alternative? Will we in the future look back nostalgically at the freedom and privacy of dumb cities? Of course happier futures are always imaginable. But urban vacation rentals are not an isolated phenomenon; they are part of evolving societal forces and ideologies that make those happier futures less plausible.

The Commodification of Everything

Airbnb and other vacation rentals are not a driver of change; they are a symptom. They are currently among the most visible manifestations of an economic evolution and its corresponding ethics that invites commercial activities to pervade and take over every aspect of our lives, even those areas for which this would have been unimaginable – or immoral – a few decades ago. This colonisation or 'enclosure' by market dynamics of domains formerly not subject to ownership transactions explains the pessimism of the scenarios, as they threaten to bring us to unprecedented levels of alienation and ongoing impoverishment, both culturally and materially.

We have sadly become accustomed to the privatisation of the public assets that used to surround us and that, ironically, we truly shared. Public transport, telecommunications – even rain water in Bolivia or sunlight in Spain – have become subject to the calculus of profitability. The same has occurred with public institutions such as education or with social welfare provisions such as health care, housing or pensions (Harvey, 2005). The chaotic deterioration of the British railway system (Meek, 2014), the inefficiencies of privatised health care (Rijk & Venema, 2016), the market failures of the Spanish housing market (Sunderland, 2014) or the post-crisis labour market (Smith, 2018), and the general abundant evidence of economic underperformance under neoliberal regimes (Harvey, 2005), have not been sufficient to convince us that free market orthodoxy serves only certain interests, but not those of resource allocation efficiency or social wellbeing.

'Touristification' is part of this same pillage, as it privatises shared cultural capital, turning it into visitor attractions and merchandise (Harvey, 2002). Locals become symbolically alienated from this 'touristified' culture, such as in the case of the liberal sex, drugs and rock 'n' roll nightlife in Amsterdam, or physically displaced, such as locals from Barcelona who avoid the Ramblas or the Boqueria market. The tragedy of mass tourism is that it must constantly seek to identify new sources of 'authenticity', which motivates people to travel, in order to subsequently commodify those sources, thus depriving them of authenticity.

The final consequence of this unsustainable development will be that tourist cities, which attract visitors with a distinct local culture, become spatially separated from the bearers of that very culture, their former residents. Even if physical 'enclosure' measures of common spaces are currently aimed at protecting locals' access from visitor pressure – the turnstiles in Venice or the closing of Parc Güell in Barcelona – what matters is that tourism has imposed a fencing of public areas that will only exacerbate this contradiction. The nature of tourism will make it impossible to keep the authentic in, and the visitors out.

It would be an ingenuous illusion to think that this 'enclosure' or commercial colonisation would stop once it had commodified everything around us. Beyond the privatisation of public assets, our personal life still contains open spaces that can – or should, according to neoliberal ethics – be conquered and turned into profitable ventures. In other words, the commodification of everything implies the *commodification of us*. To start with are our 16 daily unproductive hours: in the gig-economy, lying on your couch has an opportunity cost as you could have accepted some small paid work contract through a chores platform or Uber (Schnitzler, 2015). And while it is untrue that our homes are idle assets – since we live in them – it can be said that they are not yet economically productive. But precisely because they are not idle, that productivity comes at a cost. Can we still afford to invite friends and family to our homes if there is, for instance, a sports event in town, if that is the moment when we should be receiving tourists (Frenken, 2016)?

Profit thus can become the guiding principle of our personal lives, displacing other values like pleasure, loyalty and friendship. The so-called 'experience economy' incites entrepreneurial fantasies to commodify not only public behaviour – by managing the consumption process in a restaurant, or an airline passenger's enjoyment of his flight – but also those activities formerly considered private and intimate. That this is not just a stereotyped caricature of neoliberal greed is illustrated by the development of commercial dating platforms. After the first successes of mainstream dating platforms, a Dutch company launched the brand Second Love ('*No one has to know*'),[1] aimed at the niche market of married individuals who want to cheat on their spouses. This has apparently been a success – the platform has expanded to Belgium, Spain, Portugal and South America – and four years later the company found, with BiLove, a new niche in 'bicurious' women looking for extramarital lesbian affairs. The ethical shift provoked by neoliberal values does not refer, of course, to individual choices in sexuality, but to the commodification of intimate behaviour, and infidelity in particular.

Turning our private residence into tourist accommodation may not sound as controversial, but it similarly causes our personal sphere to become conditioned by market forces. Could it be argued that this commercialisation sets the home owner or tenant free, since he or she after all controls the residence, transformed into a means of production? This would certainly not apply to the multilisters who control between one-third and two-thirds of the units offered on Airbnb, among whom even an accumulation of capital has occurred in the most orthodox sense of the word. Neither is it true for single-listers, whose production depends on the platform even if they remain liable for financial risks. Their status as sole traders is, as so often nowadays, a disguise for flexibilised wage labour.

Among Airbnb hosts, 53% assert that they use the platform to pay their rents. If true, this does not imply, however, that Airbnb is the 'solution', as the company's European managing director argues (Whyte, 2018). But it does suggest that working as an Airbnb host, necessitated by rent hikes, is not an act of choice, regardless of whether these hikes are caused by broader urban developments rather than exclusively by the platforms. In order to keep doing what they used to do and keep living where they used to live, urban residents are dispossessed of their free time and space. In short, it would be a simplification to say that Airbnb alone has caused these problems; it would be cynical to say that it helps solve them.

From the opposite perspective, this downward middle-class mobility must be compensated for somehow to avoid consumption falling and dragging down the economic system. In this sense, urban vacation rentals and other forms of so-called 'sharing' constitute a double-edged sword: on the one hand, they are a necessary complement for stagnating wages; on the other, they contribute along with other low-cost trends to a reduction in consumer prices. The growth of tourism is made possible, in other words, as long as the urban middle classes can afford to travel thanks to falling prices, which they subsidise themselves with unregulated host labour.

The 'enclosure' and commodification of private homes is a contemporary form of enslavement, which expels us from our neighbourhoods or tolerates our presence in exchange for host labour. Cities and neighbourhoods are mined for authenticity, an illusory promise to travellers that tourism depletes wherever it is found. The Airbnb proposition denounces the fakery of mass tourism, but at the same time it colonises and commodifies the personal lives of its customers under the cynical promise of 'existential authenticity' or self-actualisation.

Future Outlook: Our Cities

The story of urban vacation rentals has no good ending, at least for those who question the ethics of neoliberalism. The phenomenon cannot escape the fundamental contradictions between alienation and authenticity, or between impoverishment and consumption growth. The 'sharing' rhetoric itself is an example of capital's capacity to incorporate, commodify and market dissenting voices and counter-cultures (Harvey, 2002; Rifkin, 2000). The word 'sharing' has thus adopted its exact opposite meaning: just as public assets are no longer shared but owned and subject to market dynamics, our personal spaces and lives have been redefined as owned commodities. They are owned by us, but our ownership rights are acknowledged with the sole purpose of enabling us to sell them for a profit.

The dreams of a micropreneurial utopia have ignored the inherent emergence of capital accumulation, which by now has come into plain sight. The anti-consumerist and environmentalist message is contradicted by the share of the platforms in the ongoing rise of tourist numbers, as well as by the similarity in tourist behaviour between 'sharing' travellers and mainstream tourists. So far, the major causes for concern in the public debate have been the negative externalities, especially the impact on cities and residents.

These third parties bear the cost of transactions in which they do not take part; in other words, market parties – platforms, providers and tourists – 'externalise' their liabilities. Pollution is the classical example of this type of market failure (Harvey, 2005), and in the case of urban vacation rentals this translates as direct nuisance, excessive use of public assets and 'touristification'. The only remedy is public regulation and control. Such regulation can only become effective if, as we have argued, rental platforms give up their secrecy.

What does this mean for the future of the tourist city? It is customary in futures studies to illustrate outlooks with scenario narratives – fictitious accounts of envisioned futures, not necessarily accurate but, rather, extreme and imaginative versions of what may happen in order to spur the discussion. In our case, this is hardly necessary. The visible manifestations of 'touristification' have become so absurd and offensive that they are beyond our wildest futurist imagination. Recent images of an idyllic sunset on Santorini show immense crowds that remind us of Times Square on New Year's Eve. The turnstiles in the pedestrian main streets of Venice or the entrance restrictions to Barcelona's Parc Güell physically undermine our fantasies of blending in with locals. On the other hand, naked tourists in the Barceloneta, Machu Picchu or Angkor Wat show little signs of such an aspiration to blend in, and rather celebrate the appropriation and commodification of these places as party zones. Amsterdam residents experience the downside of this blending in, when they are late for work because tourists who 'live like locals' keep the bathrooms of their private homes occupied.

Future Outlook: Vacation Rental Platforms

A second concern that deserves a wider debate is the tourist market itself. The market parties directly involved in 'sharing' transactions benefit, at first glance, from opportunities to make profits with their personal homes or to reduce the price of their holidays. In a sense, vacation rentals and 'sharing' in general can be considered as a privatised safety net that substitutes for reduced public welfare provisions (Morozov, 2014). But need and greed – the stick and carrot of neoliberalism – have determined the growth dynamics of short-term rentals, resulting in an accumulation of capital. Some hosts are driven to the platforms for an income that helps them make ends meet. Others generate additional income that is reinvested in an expansion of their enterprises.

It is not hard to envision how this will further evolve. With the current double- or triple-digit growth of overnight stays on Airbnb, it is likely that this accumulated capital will be invested in increasing overall supply. Once the market of tourist home 'sharing' matures and growth curves flatten, it seems plausible to predict a market concentration in which accumulated capital finds its way to existing supply. Small providers – single-listers – cannot operate with the same efficiency as multilisters and will see their revenues decrease. Those who used the platform to make some extra money may opt for leaving the business: for them, this will mean a simple impoverishment, e.g. less money to spend on holidays. As an alternative, they may outsource their operation to a professional provider, allowing them to benefit from advantages of scale.

Urban vacation rental hosts who rely on their income to pay their mortgage or their rent will not have these same options. Leaving the business will imply that they will have to give up their house; in business terms, this would be similar to a hostile take-over by a multilister. The alternative for these less affluent hosts will be to seek income security in exchange for their entrepreneurial independence. In the platform economy, it is more plausible that they will be offered facility support than traditional, risk-free employment. In other words, they become a precarious vacation rental workforce, whose productivity depends on larger organisations taking care of service consistency and sales.

Again, we do not have to be overly imaginative to understand how this will work. We have already seen how Airbnb hosts in US cities received compliance and standardisation instructions from Airbnb's San Francisco headquarters, adding to the 'hassle' and routine of receiving guests, and alienating hosts who were mainly motivated by the social contact with travellers. In most scenarios, convergence – of platforms becoming a hotel-like service and hotel companies offering residential apartments – was the most plausible evolution of urban vacation rentals. Such standardisation will also imply that the autonomy and other working conditions of Airbnb hosts will converge with those of traditional hospitality employees.

The scenario of a divergent evolution only becomes more probable if urban vacation rentals become cornered by restrictive regulations and active prosecution. Commercial operations will then most likely resort exclusively to the low-cost segment. Both in the convergence and in the divergence trend, there will be room for counter-trends of non-monetary, genuine 'sharing'; the identification of a segment driven by the ideals of collaborative consumption, about one-fifth of the total vacation rental market in size, shows the potential for such a movement.

From a broader perspective, the transformation of micropreneurial hosts into dependent employees is not an exceptional nor an isolated phenomenon. In the long run, hosts will not only lose their autonomy but also their spending power. But if we remember that a proportion of the hosts used their income to pay for leisure activities and holidays, there must be an inevitable downward spiral; after all, these hosts would be likely to use Airbnb and other platforms on their trips, so their revenue loss affects tourist demand. This will mean that to keep up their level of consumption we will have to reduce the price of production inputs. We might even come up with innovative ways to reduce the price of travelling ...

And so on.

The Future of Commodified Tourism

Concerns around environmental and cultural sustainability and the harm caused to destinations has led to calls to reduce or even stop travelling (Smith, 2018). But also from the perspective of the traveller, we might question the purpose and pleasure of commodified tourist experiences.

The contemporary mass tourist is tantalised by the promise of authenticity, which recedes whenever the visitor comes too close. Nevertheless, we have no evidence that 'touristified' cities lose their appeal. Rather, our cities keep drawing hundreds of thousands of visitors who through the 'sharing' marketing seem convinced that they are not mass tourists. Even more ironically, Airbnb has employed an anti-consumerist narrative to promote an absurd extreme of meaningless consumption. The quest for our authentic selves puts us on intercontinental flights. We need to travel to feel like a local. We want to live in Paris to sleep on a couch and to bond with our families. And, alienated from our home towns, we have the illusion of recovering a sense of belonging in a remote tourist city.

Note

(1) Secondlove.nl, apparently active since 2010 (the first blogpost is from 3 January 2010).

References

Abad Liñán, J.M. (2015) Airbnb quiere que Barcelona sea como Ámsterdam. *El País*, 4 August. See https://elpais.com/tecnologia/2015/08/04/actualidad/1438695849_391117.html (accessed 8 May 2018).

Airbnb (2012) *Study Finds that Airbnb Hosts and Guests Have Major Positive Effect on City Economies*. See https://www.airbnb.nl/press/news/study-finds-that-airbnb-hosts-and-guests-have-major-positive-effect-on-city-economies (accessed 13 October 2015).

Airbnb (2013a) *New Study: Airbnb Community Contributes €185 million to Parisian Economy*. See https://www.airbnb.nl/press/news/new-study-airbnb-community-contributes-185-million-to-parisian-economy (accessed 13 October 2015).

Airbnb (2013b) *New Study: Airbnb Community Makes Amsterdam Economy Stronger*. See https://www.airbnb.nl/press/news/new-study-airbnb-community-makes-amsterdam-economy-stronger (accessed 13 October 2015).

Airbnb (2013c) *New Study: Airbnb Generated $632 Million in Economic Activity in New York*. See https://www.airbnb.nl/press/news/new-study-airbnb-generated-632-million-in-economic-activity-in-new-york (accessed 13 October 2015).

Airbnb (2013d) *New Study: Airbnb Community Generates $824 Million in Economic Activity in the UK*. See https://www.airbnb.nl/press/news/new-study-airbnb-community-generates-824-million-in-economic-activity-in-the-uk (accessed 13 October 2015).

Airbnb (2013e) *New Study: Airbnb Community Contributes $175 Million to Barcelona's Economy*. See https://www.airbnb.nl/press/news/new-study-airbnb-community-contributes-175-million-to-barcelona-s-economy (accessed 13 October 2015).

Airbnb (2014a) *New Study: Airbnb Community Generates $61 Million in Economic Activity in Portland*. See https://www.airbnb.nl/press/news/new-study-airbnb-community-generates-61-million-in-economic-activity-in-portland (accessed 13 October 2015).

Airbnb (2014b) *New Study: Airbnb Community Generates $312 Million in Economic Impact in LA*. See https://www.airbnb.at/press/news/new-study-airbnb-community-generates-312-million-in-economic-impact-in-la (accessed 13 October 2015).

Airbnb (2014c) *New Study: Airbnb Community Generates $51 Million in Economic Impact in Boston*. See https://www.airbnb.nl/press/news/new-study-airbnb-community-generates-51-million-in-economic-impact-in-boston (accessed 13 October 2015).

Airbnb (2015a) *Airbnb Community Tops $1.15 Billion in Economic Activity in New York City*. See https://www.airbnb.com/press/news/airbnb-community-tops-1-15-billion-in-economic-activity-in-new-york-city (accessed 13 October 2015).

Airbnb (2015b) *Rónán. Meet the Hosts*. See https://www.youtube.com/watch?v=3_djW-fio5x0 (accessed 14 March 2018).

Airbnb (2016a) *Airbnb and The Rise of Millennial Travel*. See https://www.airbnbcitizen.com/wp-content/uploads/2016/08/MillennialReport.pdf (accessed 23 February 2018).

Airbnb (2016b) *All in the Family: A Study of Family Travel on Airbnb*. See https://2sqy5r1jf93u30kwzc1smfqt-wpengine.netdna-ssl.com/wp-content/uploads/2016/06/FamilyTravelReport_USA_061816_v3.pdf (accessed 23 February 2018).

Airbnb (2016c) *Jeff. Meet the Hosts.* See https://www.youtube.com/watch?v=wog_27wgKUU (accessed 14 March 2018).

Airbnb (2017) *San Francisco's Registration Process: Frequently Asked Questions.* See https://www.airbnb.com/help/article/1849/san-francisco-s-registration-process—frequently-asked-questions#How%20much%20does%20it%20cost%20to%20register (accessed 19 May 2018).

Airbnb (2018a) *Superhost: Get Recognized for Being an Outstanding Host.* See https://www.airbnb.com/superhost (accessed 19 March 2018).

Airbnb (2018b) *Airbnb is for Everyone. An Open Letter from Airbnb to Boutique Hotel and B+B Owners.* See https://www.airbnbcitizen.com/airbnb-for-everyone/ (accessed 28 March 2018).

Airbnb (n.d.) *What Are Airbnb's Standards for Hotels?* See https://www.airbnb.com/help/article/1526/what-are-airbnb-s-standards-for-hotels (accessed 30 March 2018).

Airbnb Citizen (2017) *Tax Agreements with 275 Governments.* See https://www.airbnbcitizen.com/airbnb-tax-facts/ (accessed 2 June 2018).

AirDNA (2018) *Marketminder Singapore.* See https://www.airdna.co/market-data/app/sg/default/singapore/overview (accessed 20 May 2018).

Ajuntament de Barcelona (2016) *Estadístiques de turisme. Barcelona: ciutat i entorn.* Barcelona: Ajuntament de Barcelona/Diputació de Barcelona/Barcelona Turisme.

All about Airbnb (2016) *Airbnb gets into the Property Management Business by Letting Superhosts Look After Your Place for You.* See https://all-about-airbnb.com/post/150019756336/experienced-superhosts-property-management-platform (accessed 19 March 2018).

Allor, B. (2018a) *Visitor Volume and Direct Spending Estimates, 2017.* San Francisco, CA: San Francisco Travel Association.

Allor, B. (2018b) *San Francisco City & County Lodging Statistics, 2017–Present.* San Francisco, CA: San Francisco Travel Association.

Amsterdam in Progress (2018) *Amsterdam Balanswijzer.* Amsterdam: Amsterdam in Progress.

Amsterdam Marketing (2016) *2015 in cijfers.* Amsterdam: Amsterdam Marketing.

Amsterdam Smart City (2016) *Action Plan Sharing Economy.* See https://www.slideshare.net/shareNL/amsterdam-actionplan-sharing-economy (accessed 30 April 2018).

Anderson, C.K. (2009) *The Billboard Effect: Online Travel Agent Impact on Non-OTA Reservation Volume.* Ithaca, NY: Cornell Hospitality Report.

Anderson, C.K. (2011) *Search, OTAs, and Online Booking: An Expanded Analysis of the Billboard Effect.* Ithaca, NY: Cornell Hospitality Report.

Angelidou, M. (2014) Smart city policies: A spatial approach. *Cities* 41, S3–S11.

Araya, D. (2015) *Interview: Michel Bauwens on Peer-to-peer Economics and its Role in Reshaping our World.* See http://futurism.com/interview-michel-bauwens-on-peer-to-peer-economics-and-its-role-in-reshaping-our-world/ (accessed 6 April 2016).

Arias-Sans, A. (2015) *Desmuntant Airbnb. Apunts crítics sobre el cas de Barcelona.* See http://latramaurbana.net/2015/07/01/desmuntant-airbnb-apunts-crítics-sobre-el-cas-de-barcelona/#more-1245 (accessed 31 October 2015).

Arias-Sans, A. (2018) Turisme i Gentrificació. Apunts des de Barcelona. *Papers: Regió Metropolitana de Barcelona. Territori, estratègies, Planejament* 60, 130–139.

Armstrong, M. (2006) Competition in two-sided markets. *RAND Journal of Economics* 37 (3), 668–691.

Armstrong, M. and Wright, J. (2007) Two-sided markets, competitive bottlenecks and exclusive contracts. *Economic Theory* 32 (2), 353–380.

Arran (2017) *Comunicat sobre les darreres accions de denúncia del model turístic dels Països Catalans.* See http://arran.cat/blog/2017/08/06/comunicat-sobre-darreres-accions-denuncia-model-turistic-dels-paisos-catalans/ (accessed 4 May 2018).

Associated Press (2018) Singapore court imposes hefty fines for Airbnb rentals. *CTV News*, 3 April. See https://www.ctvnews.ca/business/singapore-court-imposes-hefty-fines-for-airbnb-rentals-1.3868967 (accessed 20 May 2018).

AT5 (2018) *Is wijkverbod Airbnb juridisch haalbaar? 'Ik denk dat het mogelijk is'*. See http://www.at5.nl/artikelen/182182/is-wijkverbod-airbnb-juridisch-haalbaar-ik-denk-dat-het-mogelijk-is- (accessed 17 May 2018).

Atkin, D. (2004) *The Culting of Brands: When Customers Become True Believers*. New York: Penguin.

Atkin, D. (2014) *Community Culture*. See https://www.slideshare.net/FeverBee/douglas-atkin-how-to-create-a-powerful-community-culture (accessed 28 January 2018).

Aunión, J. and Clemente, Y. (2018) El efecto en cascada de los pisos turísticos en Madrid. *El País*, 22 February. See https://elpais.com/ccaa/2018/02/22/madrid/1519304667_585227.html (accessed 25 February 2018).

Au-Yong, R. (2018) Airbnb hosts who made illegal home rentals earned at least $19,000 over 5 weeks. *The Straits Times*, 27 February. See https://www.straitstimes.com/singapore/courts-crime/duo-who-made-illegal-home-rentals-earned-at-least-19000-over-5-weeks (accessed 23 May 2018).

Badia, E. (2013) How Airbnb saved my house. *New York Post*, 8 December. See https://nypost.com/2013/12/08/how-airbnb-saved-my-house/ (accessed 1 February 2018).

Bakker, D., Dessauvagie, M. and Oskam, J. (2016a) *Airbnb: Impacts and Outlook for Amsterdam*. Amsterdam: Colliers International and Hotelschool The Hague.

Bakker, D., Dessauvagie, M. and Oskam, J. (2016b) *Airbnb: Impact and Outlook for London*. Amsterdam: Colliers/Hotelschool The Hague.

Bakker, D., Dessauvagie, M. and Oskam, J. (2016c) *Airbnb: Impact and Outlook for Barcelona*. Amsterdam: Colliers/Hotelschool The Hague.

Bakker, D., Dessauvagie, M. and Oskam, J. (2017a) *Airbnb: Impact and Outlook for Amsterdam 2016*. Amsterdam: Colliers/Hotelschool The Hague.

Bakker, D., Dessauvagie, M. and Oskam, J. (2017b) *Airbnb in Rotterdam en Den Haag*. Amsterdam: Colliers/Hotelschool The Hague.

Bakker, D., Dessauvagie, M. and Oskam, J. (2017c) *Airbnb in Reykjavík*. Amsterdam: Colliers/Hotelschool The Hague.

Bakker, D., Dessauvagie, M. and Oskam, J. (2017d) *Airbnb in Berlin*. Amsterdam: Colliers/Hotelschool The Hague.

Bakker, D., Dessauvagie, M. and Oskam, J. (2018a) *Airbnb in Nederland: De belangrijkste cijfers over 2017*. Amsterdam: Colliers/Hotelschool The Hague.

Bakker, D., Dessauvagie, M., Harrington, D., Priesemeister, J. and Oskam, J. (2018b) *Airbnb in Europe: Key Cities Compared. Amsterdam, Berlin, London, Madrid, Paris*. Amsterdam: Colliers International/Hotelschool The Hague.

Balanzó, R. and Rodríguez-Planas, N. (2017) *Crisis and Reorganization in Urban Dynamics: The Barcelona Case Study*. IZA Discussion Paper Series No. 10748. Bonn: IZA Institute of Labor Economics.

Baquero, C. (2014) La Barceloneta estalla contra el 'turismo de borrachera' en el barrio. *El País*, 20 August. See https://elpais.com/ccaa/2014/08/20/catalunya/1408562737_343739.html (accessed 6 May 2018).

Barcelona Activa (2017) *Barcelona Data Sheet 2017. Main Economic Indicators for the Barcelona Area*. Barcelona: Barcelona City Council.

Barcelona en Comú (2015) *Programa Electoral. Municipales 2015*. See https://barcelonaencomu.cat/sites/default/files/programaencomun_cast.pdf (accessed 6 May 2018).

Bardhi, F. and Eckhardt, G.M. (2012) Access-based consumption: The case of car sharing. *Journal of Consumer Research* 39 (4), 881–898.

Barragan, B. (2016) *Anaheim Flat-out Bans Airbnb and Other Short-term Rentals*. See https://la.curbed.com/2016/6/30/12072054/anaheim-flat-out-bans-airbnb-and-short-term-rentals (accessed 4 June 2018).

Barron, K., Kung, E. and Proserpio, D. (2018) *The Sharing Economy and Housing Affordability: Evidence from Airbnb*. See https://ssrn.com/abstract=3006832 (accessed 29 June 2018).

Bauwens, M. (2005) *1000 Days of Theory: The Political Economy of Peer Production*. See www.ctheory.net/articles.asp x?id=499 (accessed 29 December 2017).

Bauwens, M. (2014) *Peer to Peer*. Amsterdam: Pakhuis De Zwijger.

Bauwens, M., Mendoza, N. and Iacomella, F. (2012) *Synthetic Overview of the Collaborative Economy*. Chiang Mai: P2P Foundation.

Beck, L. (2018) Berlin had some of the world's most restrictive rules for Airbnb rentals. Now it's loosening up. *Washington Post*, 27 March. See https://www.washingtonpost.com/world/europe/berlin-had-some-of-the-worlds-most-restrictive-rules-for-airbnb-rentals-now-its-loosening-up/2018/03/27/e3acda90-2603-11e8-a227-fd2b009466bc_story.html (accessed 28 March 2018).

Beer, T.d. and Gier, M.d. (2016) *Opinie: Airbnb, welgesteld linksliberaal tegen de 'rest'?* See http://www.duurzaambedrijfsleven.nl/retail/15675/opinie-airbnb-welgesteld-links-liberaal-tegen-de-rest (accessed 21 June 2016).

Bee Token (2017) *What is Beenest? How The Bee Token is Revolutionizing The Home Sharing Market*. See https://medium.com/@thebectoken/what-is-beenest-how-the-bee-token-is-revolutionizing-the-home-sharing-market-8da32d79bbbb (accessed 27 June 2018).

Belk, R. (2010) Sharing. *Journal of Consumer Research* 36 (5), 715–734.

Belk, R. (2014) You are what you can access: Sharing and collaborative consumption online. *Journal of Business Research* 67 (8), 1595–1600.

BeMate (2014a) *Room Mate Hotels Announces BeMate.com: New Apartment Rental And Hotel Hybrid*. See https://www.prnewswire.com/news-releases/room-mate-hotels-announces-bematecom-new-apartment-rental-and-hotel-hybrid-275589141.html (accessed 2016 March 2018).

BeMate (2014b) Brand new concept BeMate.com offering innovative new holiday apartment rentals in some of the world's most exciting cities with the best hotel services. *BusinessWire*, 17 September. See https://www.businesswire.com/news/home/20140917006150/en/Brand-New-Concept-BeMate.com-Offering-Innovative-New (accessed 26 March 2018).

Benner, K. (2017a) Inside the hotel industry's plan to combat Airbnb. *New York Times*, 16 April. See https://www.nytimes.com/2017/04/16/technology/inside-the-hotel-industrys-plan-to-combat-airbnb.html (accessed 21 March 2018).

Benner, K. (2017b) Airbnb tries to behave more like a hotel. *New York Times*, 17 June. See https://www.nytimes.com/2017/06/17/technology/airbnbs-hosts-professional-hotels.html (accessed 29 June 2017).

Berg, J. (2016) *Income Security in the On-demand Economy: Findings and Policy Lessons from a Survey of Crowdworkers*. Geneva: International Labour Office.

Bergman, J. (2017) Why Airbnb is well placed to crack the China market. *The Guardian*, 31 March. See http://www.theguardian.com/travel/2017/mar/31/why-airbnb-well-placed-to-crack-china-holiday-rental-market-aibiying (accessed 12 January 2018).

BeWelcome (2018) *BeWelcome Statistics*. See https://www.bewelcome.org/stats (accessed 18 May 2018).

Billock, J. (2015) *Ask the CEO: Revenue at Couchsurfing*. See https://blog.couchsurfing.com/ask-the-ceo-revenue-at-couchsurfing/ (accessed 18 May 2018).

Bishop, P., Hines, A. and Collins, T. (2007) The current state of scenario development: An overview of techniques. *Foresight* 9 (1), 5–25.

Blickhan, M., Bürk, T. and Grube, N. (2014) Touristification in Berlin. *Sub\urban. Zeitschrift für kritische Stadtforschung* 2 (1), 167–180.

Bloomberg (2018) Airbnb to share information with authorities on guests in China. *Bloomberg*, 29 March. See https://www.bloomberg.com/news/articles/2018-03-29/airbnb-to-share-information-with-authorities-on-guests-in-china (accessed 8 June 2018).

Board of Supervisors (2014) *President Chiu Announced Legislation to Regulate Short-Term Rentals and Protect Residential Housing*. See https://sfbos.org/

president-chiu-announced-legislation-regulate-short-term-rentals-and-protect-residential-housing (accessed 14 May 2018).

Booth, K. and Kiss, J. (2015) San Francisco voters reject proposition to restrict Airbnb rentals. *The Guardian*, 4 November. See https://www.theguardian.com/us-news/2015/nov/04/san-francisco-voters-reject-proposition-f-restrict-airbnb-rentals (accessed 3 January 2018).

Bosa, D. (2017) Airbnb lashes out at Marriott as clash between Silicon Valley and the hotel industry intensifies. *CNBC*, 20 November. See https://www.cnbc.com/2017/11/20/airbnb-wrote-a-letter-to-marriott-claiming-hotel-fleeces-taxpayers.html (accessed 22 November 2017).

Bostoen, F. (2018) Neutrality, fairness or freedom? Principles for platform regulation. *Internet Policy Review* 7 (1), 2–19.

Botsman, R. (2010) *The Case for Collaborative Consumption*. See http://www.ted.com/talks/rachel_botsman_the_case_for_collaborative_consumption/ (accessed 29 October).

Botsman, R. and Rogers, R. (2011) *What's Mine is Yours: How Collaborative Consumption is Changing the Way We Live*. London: Harper Collins.

Bouma, F. and Middel, M. (2018) *Amsterdam wil Airbnb in toeristische buurten verbieden*. See https://www.nrc.nl/nieuws/2018/05/16/amsterdam-verbiedt-airbnb-in-toeristische-buurten-a1603106 (accessed 7 June 2018).

Bowers, B. (2017) Airbnb in Paris. Global Manager Abroad No. 4. Lewisburg, PA: Bucknell Digital Commons.

Bowles, N. (2018) Dorm living for professionals comes to San Francisco. *New York Times*, 4 March. See https://www.nytimes.com/2018/03/04/technology/dorm-living-grown-ups-san-francisco.html (accessed 15 May 2018).

B.R. (2016) Business travellers will be hit hardest by the crackdown on Airbnb. *The Economist*, 4 November. See https://www.economist.com/blogs/gulliver/2016/11/lets-down (accessed 4 March 2018).

Brinklow, A. (2018) *Travel Association Blames Homeless Crisis for Stalling San Francisco Tourism*. See https://sf.curbed.com/2018/2/23/17046028/travel-association-tourism-homeless-trump (accessed 15 May 2018).

Brousseau, F. (2016) *Policy Analysis Report to Supervisor Campos Re: Short-Term Rentals 2016 Update*. San Francisco, CA: Budget and Legislative Analyst's Office.

Brunton, J. (2018) Venice poised to segregate tourists as city braces itself for May Day 'invasion'. *The Guardian*, 1 May. See https://www.theguardian.com/travel/2018/may/01/venice-to-segregate-tourists-in-may-day-overcrowding (accessed 2 May 2018).

Buhalis, D. and Licata, M.C. (2002) The future eTourism intermediaries. *Tourism Management* 23, 207–220.

Business Travel IQ (2017) *Is Brussels Back in Business?* See https://www.businesstravel-iq.com/article/2017/05/04/is-brussels-back-in-business (accessed 29 June, 2018).

Caillaud, B. and Jullien, B. (2003) Chicken & egg: Competition among intermediation service providers. *RAND Journal of Economics* 34 (2), 309–328.

Calder, S. (2018) How the home-stay rental network will change 2018. *The Independent*, 23 February. See https://www.independent.co.uk/travel/news-and-advice/airbnb-changes-2018-how-upgraded-plus-properties-hotels-b-bs-a8225046.html (accessed 28 March 2018).

Carey, M. (2018a) *Airbnb Plus Is for People Who Hate Airbnb*. See https://www.cntraveler.com/story/airbnb-plus-is-for-people-who-hate-airbnb (accessed 28 March 2018).

Carey, M. (2018b) *Spain Cracked Down Hard on Airbnb Last Week*. See https://www.cntraveler.com/story/spain-cracked-down-big-on-airbnb-last-week (accessed 17 May 2018).

Carey, M. (2018c) *Valencia Doesn't Want You Renting an Airbnb with a View*. See https://www.cntraveler.com/story/valencia-doesnt-want-you-renting-an-airbnb-with-a-view (accessed 7 June 2018).

Carey, M. (2018d) *Nearly 80 Percent of Japan's Airbnbs Were Just Removed.* See https://www.cntraveler.com/story/nearly-80-percent-of-japans-airbnbs-were-just-removed (accessed 5 June 2018).

Carey, M. (2018e) *Paris Could Pull 43,000 Airbnb Listings by This June.* See https://www.cntraveler.com/story/paris-could-pull-43000-airbnb-listings-by-this-june (accessed 8 June 2018).

Carroll, B. and Siguaw, J. (2003) *Evolution in Electronic Distribution: Effects on Hotels and Intermediaries.* Ithaca, NY: Center For Hospitality Research, Cornell University.

Carson, B. (2017) *Airbnb's New Apartment Brand Lands $200 Million Investment.* See https://www.forbes.com/sites/bizcarson/2017/12/18/airbnbs-apartment-brand-niido-investment/#416f42075049 (accessed 26 March 2018).

Castillo, I.A. (2017a) *El Centro de Málaga vive un boom económico pero pierde 200 vecinos en dos años.* See http://www.laopiniondemalaga.es/malaga/2017/04/29/centro-malaga-vive-boom-economico/926952.html (accessed 22 April 2018).

Castillo, I.A. (2017b) *Los pisos turísticos inflan la burbuja de los alquileres con precios récord y poca oferta.* See http://www.laopiniondemalaga.es/malaga/2017/11/27/pisos-turisticos-inflan-burbuja-alquileres/970558.html (accessed 22 April 2018).

Catà, J. (2017) Airbnb: 'El Ayuntamiento prefiere el conflicto ante el acuerdo'. *El País,* 9 May. See https://elpais.com/ccaa/2017/05/09/catalunya/1494342628_146451.html (accessed 8 May 2018).

Cave, B. (2016) *Airbnb Takes More Conciliatory Approach Following Adverse Ruling.* See https://www.lexology.com/library/detail.aspx?g=4d95164b-3c43-4283-a201-413b86d69145 (accessed 14 May 2018).

CEIBS (2017) *Xiaozhu.com: A Case Study on the Sharing Economy.* See http://www.ceibs.edu/xiaozhu.com-case-study-sharing-economy (accessed 12 January 2018).

Channel News Asia (2017) *Minimum Rental Period for Private Homes Cut to 3 Months.* See https://www.channelnewsasia.com/news/singapore/minimum-rental-period-for-private-homes-cut-to-3-months-8991156 (accessed 23 May 2018).

Chernyshenko, O.S., Uy, M.A., Jiang, W., Ho, M.H.R., Lee, S.P., Chan, K.Y. and Yu, K.Y.T. (2015) *Global Entrepreneurship Monitor 2014 Singapore Report.* Singapore: Nanyang Technological University.

Chiarelli, N. and Warriner, J. (2015) *Future Traveller Tribes 2030: Understanding Tomorrow's Traveller.* London: Amadeus.

Christides, G. (2018) Wenn sich eine Lehrerin keine Wohnung leisten kann. *Spiegel Online,* 17 April. See http://www.spiegel.de/wirtschaft/soziales/airbnb-unterkuenfte-verschaerfen-wohnungsnot-in-griechenland-a-1201185.html (accessed 22 April 2018).

City Attorney of San Francisco (2016) *Federal Judge Gives Initial Backing to SF Law that Holds Online Rental Platform Companies Accountable for Illegal Rentals.* See https://www.sfcityattorney.org/2016/11/08/federal-judge-gives-initial-backing-sf-law-holds-online-rental-platform-companies-accountable-illegal-rentals/ (accessed 14 May 2018).

City Attorney of San Francisco (2017) *Herrera Repels Legal Challenge to Short-term Rental Law, Secures Settlement with Airbnb and HomeAway.* See https://www.sfcityattorney.org/2017/05/01/herrera-repels-legal-challenge-short-term-rental-law-secures-settlement-airbnb-homeaway/ (accessed 15 May 2018).

Ciutat per a qui l'habita (2017) *Manifest.* See https://ciutatperaquilhabitablist.blog/prova/ (accessed 4 April 2018).

Clark, P. (2016) *Study: Airbnb Super Hosts Generate 39% of Revenue in Major Markets.* See https://skift.com/2016/01/20/study-airbnb-super-hosts-generate-39-of-revenue-in-major-markets/ (accessed 20 January 2016).

Coca, N. (2013) *The Rise and Fall of Couchsurfing.* See https://www.nithincoca.com/2013/03/27/the-rise-and-fall-of-couchsurfing/ (accessed 18 May 2018).

Cócola Gant, A. (2015) Tourism and commercial gentrification. Paper presented at the *RC21 International Conference on The Ideal City: Between Myth and Reality*, Urbino, Italy, 27–29 August.

Cohen, L. (2016) *Airbnb and Municipal Zoning Regulation*. See https://hostcompliance.com/resources-gallery/?offset=1461220860424 (accessed 3 June 2018).

Consigli, M. Gallagher, M., Kumar, M., Mehta, N., Purnell, J. and Templeton, R. (2012) *EIS Final Project: Airbnb*. Hanover: Dartmouth College.

Corbett, S. (2015) *Meet the Unlikely Airbnb Hosts of Japan*. New York Times, 22 February. See https://www.nytimes.com/2015/02/22/magazine/meet-the-unlikely-airbnb-hosts-of-japan.html (accessed 15 March 2018).

Couchsurfing (2017) *About Us*. See http://www.couchsurfing.com/about/about-us/ (accessed 18 May 2018).

Cox, M. (2015) *Amsterdam*. See http://insideairbnb.com/amsterdam/ (accessed 31 October 2015).

Cox, M. (2016) *NYC: Report on the Anti-Airbnb Advertising Law*. See http://insideairbnb.com/nyc-report-on-the-anti-airbnb-advertising-law/ (accessed 6 June 2018).

Cox, M. (2017a) *The Face of Airbnb, New York City. Airbnb as a Racial Gentrification Tool*. New York: Inside Airbnb.

Cox, M. (2017b) Email.

Cox, M. and Slee, T. (2016) *How Airbnb's Data Hid the Facts in New York City*. New York: Inside Airbnb.

Custer, C. (2016) *Why Airbnb is struggling in China*. See https://www.techinasia.com/airbnb-struggling-china (accessed 12 January 2018).

Cyber Report (2016) *Airbnb Challenges Santa Monica Short-Term Rental Law*. See https://ilccyberreport.wordpress.com/2016/09/14/airbnb-challenges-santa-monica-short-term-rental-law/ (accessed 6 June 2018).

Daniels, J. (2018) *Lie-detecting Computer Kiosks Equipped with Artificial Intelligence Look Like the Future of Border Security*. See https://www.cnbc.com/2018/05/15/lie-detectors-with-artificial-intelligence-are-future-of-border-security.html (accessed 16 May 2018).

Dator, J. (2009) Alternative futures at the Manoa School. *Journal of Futures Studies* 14 (2), 1–18.

Davies, S. and Mishkin, S. (2014) Airbnb in deal with Amsterdam to collect tourist taxes. *Financial Times*, 18 December. See http://www.ft.com/cms/s/0/ee07a86c-86b0-11e4-8a51-00144feabdc0.html (accessed 7 April 2016).

De Groen, W.P., Maselli, I. and Fabo, B. (2016) *The Digital Market for Local Services: A One-night Stand for Workers? An Example from the On-demand Economy*. CEPS Special Report No. 133. Brussels: Centre for European Policy Studies.

Dekker, D.M. (2014) Personality and hospitable behavior. In I. Pantelidis (ed.) *The Routledge Handbook of Hospitality Management* (pp. 57–66). New York: Routledge.

Dekker, D.M. (2018) Genuinely hospitable behavior in education. In J.A. Oskam, D.M. Dekker and K. Wiegerink (eds) *Innovation in Hospitality Education: Anticipating the Educational Needs of a Changing Profession* (pp. 65–75). Cham: Springer.

Delgado-Medrano, H.M. and Lyon, K. (2016) *Short Changing New York City. The Impact of Airbnb on New York City's Housing Market*. New York: BJH Advisors.

Deloitte (2018) *2018 Deloitte Millennial Survey. Millennials Disappointed in Business, Unprepared for Industry 4.0*. New York: Deloitte.

Demurtas, A. (2017) *90% Airbnb Increase: 44% of Rental Market Listed*. See https://grapevine.is/news/2017/05/31/90-airbnb-increase-44-of-rental-market-listed/ (accessed 1 June 2017).

Denyer, S. (2018) China's watchful eye. *Washington Post*, 7 January. See https://www.washingtonpost.com/news/world/wp/2018/01/07/feature/in-china-facial-recognition-is-sharp-end-of-a-drive-for-total-surveillance/?utm_term=.075e45e31c15 (accessed 26 May 2018).

Diário da República (2018) Lei n.º 62/2018 de 22 de agosto. *Diário da República* 161, 4300–4312.

Dillahunt, T. and Malone, A. (2015) The promise of the sharing economy among disadvantaged communities. In *Proceedings of the 33rd Annual ACM Conference on Human Factors in Computing Systems* (pp. 2285–2294). Seoul: ACM.

Dobbins, J. (2017) Make a living using Airbnb. *New York Times*, 7 April. See https://www.nytimes.com/2017/04/07/realestate/making-a-living-with-airbnb.html (accessed 15 March 2018).

Dolnicar, S. (2018) *Peer-to-Peer Accommodation Networks: Pushing the Boundaries*. Oxford: Goodfellow.

Donato, J. (2016) *Order RE Preliminary Injunction*. San Francisco, CA: United States District Court, Northern District of California.

Dudás, G., Boros, L., Kovalcsik, T. and Kovalcsik, B. (2017) The visualisation of the spatiality of Airbnb in Budapest using 3-band raster represeantation. *Geographia Technica* 12 (1), 23–30.

Duran, P. (2005) *The Impact of the Games on Tourism. Barcelona: The Legacy of the Games, 1992–2002*. See http://tourisminsights.info/ONLINEPUB/OLYMPICS/OLYMPICS%20PDFS/Duran.pdf (accessed 5 May 2018).

Eckhardt, G.M. and Bardhi, F. (2015) The sharing economy isn't about sharing at all. *Harvard Business Review*, 28 January.

Edelman, B. (2017) Uber can't be fixed – it's time for regulators to shut it down. *Harvard Business Review*, 21 June. See https://hbr.org/2017/06/uber-cant-be-fixed-its-time-for-regulators-to-shut-it-down (accessed 22 June 2017).

Edelman, B. and Geradin, D. (2015) Efficiencies and regulatory shortcuts: How should we regulate companies like Airbnb and Uber? *Stanford Technology Law Review* 19, 293.

Edelman, B.G. and Luca, M. (2014) Digital discrimination: The case of airbnb.com. Harvard Business School NOM Unit Working Paper No. 14-054.

Edwards, J. (2016) *Here's Exactly What Airbnb Does to Rent in Popular Cities*. See http://uk.businessinsider.com/statistics-data-airbnb-rent-prices-2016-10 (accessed 23 April 2018).

Eisenmann, T., Parker, G. and Alstyne, V.M.W. (2006) Strategies for two-sided markets. *Harvard Business Review* 84 (10).

El Economista (2017) *Barcelona formaliza la multa a Airbnb y le pondrá otra de 600.000 euros si sigue ofertando pisos ilegales*. See http://www.eleconomista.es/empresas-finanzas/noticias/8456522/06/17/Barcelona-formaliza-la-multa-a-Airbnb-y-le-pondra-otra-de-600000-euros-si-sigue-ofertando-pisos-ilegales.html (accessed 8 May 2018).

El Mundo (2015) Barcelona condonará la multa a apartamentos ilegales si van a alquiler social. *El Mundo*, 5 August. See http://www.elmundo.es/cataluna/2015/08/05/55c1d78d46163f32678b45ab.html (accessed 18 October 2015).

El País (2016) Barcelona, la piedra en el zapato de Airbnb. Así negociaron otras ciudades. *El País*, 26 November. See https://elpais.com/economia/2016/11/25/actualidad/1480093475_356500.html (accessed 8 May 2018).

El País (2018) La OMT advierte de la caída del turismo en Cataluña. *El País*, 10 January. See https://elpais.com/economia/2018/01/10/actualidad/1515579164_349235.html (accessed 7 May 2018).

Emprendedores (2009) *Claves del éxito de Room Mate*. See http://www.emprendedores.es/casos-de-exito/entrevista-a-kike-sarasola-room-mate-claves-exito-hoteles (accessed 25 March 2018).

Ert, E., Fleischer, A. and Magen, N. (2016) Trust and reputation in the sharing economy: The role of personal photos in Airbnb. *Tourism Management* 55, 62–73.

Espinosa, T. (2016) The cost of sharing and the common law: How to address the negative externalities of home-sharing. *Chapman Law Review* 19 (2), 597–627.

Evans, D. (2009) Background note. In *OECD Policy Roundtables: Two-sided Markets* (pp. 23–48). Paris: OECD.

Evans, D.S. and Schmalensee, R. (2016) *Matchmakers: The New Economics of Multisided Platforms*. Watertown, MA: Harvard Business Review Press.

EY España (2015) *Impactos derivados del exponencial crecimiento de los alojamientos turísticos en viviendas de alquiler en España, impulsado por los modelos y plataformas ce comercialización P2P*. Madrid: Exceltur.

Fedorova, T., Schmitz, D. and De Graaff, L. (2017) *Toerisme MRA 2016–2017*. Amsterdam: Gemeente Amsterdam OIS.

Forgacs, G. and Dimanche, F. (2016) Revenue challenges for hotels in the sharing economy: Facing the Airbnb menace. *Journal of Revenue and Pricing Management* 15 (6), 509–515.

Fradkin, A., Grewal, E., Holtz, D. and Pearson, M. (2015) Bias and reciprocity in online reviews: Evidence from field experiments on airbnb. In *Proceedings of the Sixteenth ACM Conference on Economics and Computation* (p. 641). Portland, OR: ACM.

France 24 (2015) *Airbnb Vows to Comply with Paris Lodgings Tax*. See http://www.france24.com/en/20150226-paris-airbnb-cooperate-home-rental-hotels-chesky-france (accessed 7 April 2016).

Frankel, M. (2017) *What's the Average American's Tax Rate?* See https://www.fool.com/retirement/2017/03/04/whats-the-average-americans-tax-rate.aspx (accessed 19 May 2018).

Frenken, K. (2016) *Deeleconomie onder één noemer: Inaugurele rede*. Utrecht: Utrecht University.

Füller, H. and Michel, B. (2014) 'Stop being a tourist!' New dynamics of urban tourism in Berlin-Kreuzberg. *International Journal of Urban and Regional Research* 38 (4), 1304–1318.

Gansky, L. (2010) *Why the Future is Sharing*. New York: Penguin Group.

Garcés, E. (2018) Data dollection in online platform businesses: A perspective for antitrust assessment. *CPI Antitrust Chronicle,* May, 2–9.

Garfield, L. (2016) *A New Hotel is Like WeWork Combined with Airbnb – Take a Look Inside*. See https://www.businessinsider.nl/zoku-hotel-amsterdam-us-2016-10/?international=true&r=US (accessed 25 March 2019).

Garrett-Price, K. (2014) *Friends I Made Using Airbnb*. See https://medium.com/thsppl/friends-i-made-using-airbnb-2f770d85e7c6 (accessed 18 September 2015).

Gebbia, J. (2016) *How Airbnb Designs for Trust*. See https://www.ted.com/talks/joe_gebbia_how_airbnb_designs_for_trust/transcript?language=en#t-751180 (accessed 6 April 2016).

Gibbs, C., Guttentag, D., Gretzel, U. and Morton, J. (2018) Pricing in the sharing economy: A hedonic pricing model applied to Airbnb listings. *Journal of Travel & Tourism Marketing* 35, 46–56.

Gilheany, J., Wang, D. and Xi, S. (2015) *The Model Minority? Not on Airbnb.com: A Hedonic Pricing Model to Quantify Racial Bias against Asian Americans*. See https://techscience.org/a/2015090104 (accessed 3 September 2017).

Gilmore, J.H. and Pine, J. (2007) *Authenticity*. Boston, MA: Harvard Business Review Press.

Glaeser, E.L. and Gyourko, J. (2002) The impact of zoning on housing affordability. National Bureau of Economic Research Working Paper Series No. 8835.

Glaeser, E.L., Gyourko, J. and Skas, R.E. (2005a) Urban growth and housing supply. National Bureau of Economic Research Working Paper Series No. 11097.

Glaeser, E.L., Gyourko, J. and Skas, R.E. (2005b) Why have housing prices gone up? National Bureau of Research Working Paper Series No. 11129.

Glass, R. (1964) *London: Aspects of Change, Vol. 3*. London: MacGibbon & Kee.

Goffman, E. (1959) *The Presentation of Self in Everyday Life*. Garden City, NY: Doubleday.

Goldman, A. (2013) *The 'Google Shuttle Effect': Gentrification and San Francisco's dot.com Boom 2.0*. Berkeley, CA: University of California.

Goodwin, T. (2015) *The Battle Is for the Customer Interface*. See https://techcrunch.com/2015/03/03/in-the-age-of-disintermediation-the-battle-is-all-for-the-customer-interface/ (accessed 6 January 2018).

Gravari-Barbas, M. and Jacquot, S. (2017) No conflict? Discourses and management of tourism-related tensions in Paris. In C. Colomb and J. Novy (eds) *Protest and Resistance in the Tourist City* (pp. 31–51). Abingdon/New York: Routledge.

Green, M. (2002) Napster opens Pandora's box: Examining how file-sharing services threaten the enforcement of copyright on the internet. *Ohio State Law Journal* 63 (2), 799–831.

Griswold, A. (2015) *Airbnb has a Really Convoluted Process for Being 'Transparent' with New York City*. See https://qz.com/571165/airbnb-is-sharing-more-data-but-in-a-really-convoluted-way/ (accessed 8 June 2018).

Guest to Guest (2018) *Press: The Key Figures and Trends on Home Exchange*. See https://www.guesttoguest.com/en/p/press (accessed 18 May 2018).

Gunter, U. (2018) What makes an Airbnb host a superhost? Empirical evidence from San Francisco and the Bay Area. *Tourism Management* 66, 26–37.

Gunter, U. and Önder, I. (2018) Determinants of Airbnb demand in Vienna and their implications for the traditional accommodation industry. *Tourism Economics* 24 (3), 270–293.

Gutiérrez, J., García-Palomares, J., Romanillos, G. and Salas-Olmedo, M. (2017) The eruption of Airbnb in tourist cities: Comparing spatial patterns of hotels and peer-to-peer accommodation in Barcelona. *Tourism Management* 62, 278–291.

Guttentag, D.A. (2015) Airbnb: Disruptive innovation and the rise of an informal tourism accommodation sector. *Current Issues in Tourism* 18 (12), 1192–1217.

Guttentag, D.A. (2016) Why tourists choose Airbnb: A motivation-based segmentation study underpinned by innovation concepts. DPhil thesis, University of Waterloo.

Guttentag, D., Smith, S., Potwarka, L. and Havitz, M. (2018) Why tourists choose Airbnb: A motivation-based segmentation study. *Journal of Travel Research* 57 (3), 342–359.

Gyimóthy, S. and Dredge, D. (2017) Definitions and mapping the landscape in the collaborative economy. In S. Gyimóthy and D. Dredge (eds) *Collaborative Economy and Tourism* (pp. 15–30). Cham: Springer.

Hajibaba, H. and Dolnicar, S. (2017) Airbnb and its competitors. In S. Dolnicar (ed.) *Peer-to-Peer Accommodation Networks: Pushing the Boundaries* (pp. 63–76). Oxford: Goodfellow.

Hall, C.M. (2007) Response to Yeoman *et al*: The fakery of 'The authentic tourist'. *Tourism Management* 28 (4), 1139–1140.

Hamari, J., Sjöklint, M. and Ukkonen, A. (2016) The sharing economy: Why people participate in collaborative consumption. *Journal of the Association for Information Science and Technology* 67 (9), 2047–2059.

Hantman, D. (2014) *San Francisco, Taxes and the Airbnb Community*. See https://www.airbnbcitizen.com/san-francisco-taxes-and-the-airbnb-community/ (accessed 14 May 2018).

Harboe Sorensen, V. (2016) Airbnb property management companies in the Netherlands. Unpublished BA dissertation, Hotelschool The Hague.

Hartung, R. (2014) *Why Airbnb's Asia HQ is in a Country Where it Can't Offer Rooms*. See https://www.singaporebusiness.com/2014/why-airbnb-asia-hq-is-in-a-country-where-it-cant-offer-rooms.html?subFormJan2018=visible (accessed 20 May 2018).

Harvey, D. (2002) The art of rent: Globalisation, monopoly and the commodification of culture. *Socialist Register* 38, 93–110.

Harvey, D. (2005) *A Brief History of Neoliberalism*. Oxford: Oxford University Press.

Haywood, J., Mayock, P., Freitag, J., Owoo, K.A. and Fiorilla, B. (2016) *Airbnb & Hotel Performance: An Analysis of Proprietary Data in 13 Global Markets*. Hendersonville, TN: Smith Travel Research.

Heckman, M. (2016) The strategic use of patents in standardization in relation to US, European and Chinese competition law. PhD dissertation, Maastricht University.

Heng, J. and Jo, Y.S. (2016) *Can Singapore Make Room and Rules for Airbnb and Other Home-sharing Offerings?* See https://www.straitstimes.com/singapore/housing/can-singapore-make-room-and-rules-for-airbnb-and-other-home-sharing-offerings (accessed 20 May 2018).

Hernández Cordero, A. (2015) *En transformación. Gentrificación en el Casc Antic de Barcelona.* Barcelona: Autonomous University of Barcelona.

Hill, D. (2015) *The Secret of Airbnb's Pricing Algorithm.* See http://spectrum.ieee.org/computing/software/the-secret-of-airbnbs-pricing-algorithm (accessed 26 August 2015).

Hines, A. and Bishop, P.C. (2013) Framework foresight: Exploring futures the Houston way. *Futures* 51, 31–54.

Hobsbawm, E. and Ranger, T. (1983) *The Invention of Tradition.* Cambridge: Cambridge University Press.

Hollands, R.G. (2008) Will the real smart city please stand up? *City* 12 (3), 303–320.

Hollands, R.G. (2015) Critical interventions into the corporate smart city. *Cambridge Journal of Regions, Economy and Society* 8, 61–77.

Hong, J. (2018) *Airbnb Listings Up 11% Since S'pore Ban Took Effect Last May.* See https://www.straitstimes.com/singapore/airbnb-listings-up-11-since-spore-ban-took-effect-last-may (accessed 23 May 2018).

Honigsberg, P. (2002) The evolution and revolution of Napster. *University of San Francisco Law Review* 36, 473–508.

Horwath (2017) *HOSTA.* Amsterdam: Horwath.

Host Compliance (2018) *Short-Term Vacation Rental Bans Gone Wrong.* See https://hostcompliance.com/short-term-vacation-rentals-bans-gone-wrong (accessed 6 June 2018).

Hosteltur (2018a) *Multa de 300.000 € a Airbnb por incumplir la ley balear.* See https://www.hosteltur.com/126697_multa-300000-airbnb-incumplir-ley-balear.html (accessed 19 February 2018).

Hosteltur (2018b) *Airbnb dará datos de los anfitriones al Ayuntamiento de Barcelona.* See https://www.hosteltur.com/128348_airbnb-dara-datos-anfitriones-al-ayuntamiento-barcelona.html (accessed 28 May 2018).

HOTREC (2015) *Levelling the Playing Field.* Brussels: HOTREC.

Huet, E. (2014) *How Uber's Shady Firing Policy Could Backfire On The Company.* See https://www.forbes.com/sites/ellenhuet/2014/10/30/uber-driver-firing-policy/ (accessed 1 September 2018).

Huete, R. and Mantecón, A. (2018) El auge de la turismofobia ¿hipótesis de investigación o ruido ideológico? *Pasos. Revista de Turismo y Patrimonio Cultural* 16 (1), 9–19.

Hurst, N. (2012) *Take That, Tokyo! San Francisco Approves 220-square-foot 'Micro Apartments'.* See https://www.wired.com/2012/11/san-francisco-micro-apartments/ (accessed 15 May 2018).

Iansiti, M. and Lakhani, K. (2017) The truth about blockchain. *Harvard Business Review* 95 (1), 118–127.

IDA (2011) *E-Government Masterplan 2011–2015. Collaborative Government.* Singapore: Info-communications Development Authority.

Ikkala, T. and Lampinen, A. (2014) *Defining the Price of Hospitality: Networked Hospitality Exchange via Airbnb* (pp. 173–176). Baltimore, MD: ACM.

Ikkala, T. and Lampinen, A. (2015) *Monetizing Network Hospitality: Hospitality and Sociability in the Context of Airbnb* (pp. 1033–1044). Vancouver: ACM.

Ioannides, D., Röslmaier, M. and Van der Zee, E. (2018) Airbnb as an instigator of 'tourism bubble' expansion in Utrecht's Lombok neighbourhood. *Tourism Geographies,* April.

IPK (2015) *ITB World Travel Trends Report 2015–2016.* Berlin: Messe Berlin GmbH.

Jacobs, T. (2016) *Interview Laurens Ivens over Airbnb: 'Hoteliers moeten minder klagen'*. See https://www.hospitality-management.nl/interview-laurens-ivens-over-airbnb-hoteliers-moeten-minder-klagen (accessed 8 May 2018).

JLL (2017) *Hotel Destinations Asia Pacific*. Chicago, IL: JLL Hotels & Hospitality Group.

Johnson, B. (2011) *After Going for-profit, CouchSurfing Faces User Revolt*. See https://gigaom.com/2011/09/01/after-going-for-profit-couchsurfing-faces-user-revolt/ (accessed 18 May 2018).

Joo, D. (2016) *Airbnb v. Schneiderman: Airbnb Challenges New York Law Regulating Short-term Rentals*. See https://jolt.law.harvard.edu/digest/airbnb-challenges-new-york-law-regulating-short-term-rentals (accessed 6 June 2018).

Jun Sen, N. (2017) *Two Airbnb Hosts Charged over Illegal Home Sharing in First Case under New Laws*. See https://www.straitstimes.com/singapore/two-airbnb-hosts-charged-over-illegal-home-sharing-in-first-case-under-new-laws (accessed 20 May 2018).

Juniper Research (2018) *Smart Cities – What's in it for Citizens?* Basingstoke: Juniper Research.

Kagermeier, A., Köller, J. and Stors, N. (2015) Airbnb als Share Economy-Herausforderung für Berlin und die Reaktionen der Hotelbranche. *Studien zur Freizeit- und Tourismusforschung* 11.

Kakar, V., Voelz, J., Wu, J. and Franco, J. (2017) The visible host: Does race guide Airbnb rental rates in San Francisco? *Journal of Housing Economics* 40, 25–40.

Kam Fung So, K., Oh, H. and Min, S. (2018) Motivations and constraints of Airbnb consumers: Findings from a mixed-methods approach. *Tourism Management* 67, 224–236.

Katz, L. (2012) *Interview with Emerging Hotelier Kike Sarasola of Room Mate Hotels*. See http://www.justluxe.com/luxe-insider/trends/feature-1782624.php (accessed 25 March 2018).

Khoo, L. (2016) *Short-term Rental: It's in the Phrasing of Contracts*. See https://www.businesstimes.com.sg/real-estate/short-term-rental-its-in-the-phrasing-of-contracts (accessed 23 May 2018).

Kleinherenbrink, A. (2016) A study of Airbnb and guest loyalty. Unpublished BA dissertation, Hotelschool The Hague.

Kloosterboer, D. (2017) *De afspraak tussen Amsterdam en Airbnb: een verkennende analyse met open data*. See https://dirkmjk.nl/nl/2017/06/de-afspraak-tussen-amsterdam-en-airbnb-een-verkennende-analyse-met-open-data (accessed 6 June 2018).

Kolar, T. and Zabkar, V. (2010) A consumer-based model of authenticity: An oxymoron or the foundation of cultural heritage marketing? *Tourism Management* 31 (5), 652–664.

Koomans, M. (2018) *Price Determinants of Airbnb in Rotterdam*. The Hague: Hotelschool The Hague.

Kostakis, V. and Bauwens, M. (2014) *Network Society and Future Scenarios for a Collaborative Economy*. Basingstoke and New York: Palgrave Macmillan.

Kraniotis, L. (2016) *Amsterdam gaat illegale Airbnb's aanpakken met 'big data'*. See http://nos.nl/artikel/2087106-amsterdam-gaat-illegale-airbnb-s-aanpakken-met-big-data.html (accessed 6 April 2016).

KRO-NCRV (2017) *Jan Paternotte en Sander Schimmelpenninck over de voors en tegens van Airbnb*. See https://evajinek.kro-ncrv.nl/artikelen/jan-paternotte-en-sander-schimmelpenninck-over-de-voors-en-tegens-van-airbnb (accessed 1 May 2018).

Kusisto, L. (2015) *Airbnb Pushes Up Rents Slightly, Study Says*. See https://blogs.wsj.com/developments/2015/03/30/airbnb-pushes-up-apartment-rents-slightly-study-says/ (accessed 26 March 2016).

Lalicic, L. and Weismayer, C. (2017) The role of authenticity in Airbnb experiences. In R. Schegg and B. Stangl (eds) *Information and Communication Technologies in Tourism* (pp. 781–794). Cham: Springer.

Langford, G., Weissenberg, A. and Pingitore, G. (2017) *2017 Travel and Hospitality Industry Outlook*. New York: Deloitte Center for Industry Insights.

La Vanguardia (2016) *Barcelona multa a Airbnb y Homeway con 600.000 euros por seguir anunciando pisos sin licencia*. See http://www.lavanguardia.com/local/barcelona/20161124/412132887490/barcelona-multa-airbnb-homeway-pisos-sin-licencia.html (accessed 7 May 2018).

Lee, D. (2016) How Airbnb short-term rentals exacerbate Los Angeles's affordable housing crisis: Analysis and policy recommendations. *Harvard Law & Policy Review* 10, 229–253.

Lee, T. (2013) *Singapore an Advanced Surveillance State, but Citizens Don't Mind*. See https://www.techinasia.com/singapore-advanced-surveillance-state-citizens-mind (accessed 26 May 2018).

Lees, L., Slater, T. and Wyly, E. (2013) *Gentrification*. New York: Routledge.

Lehr, D.D. (2015) An analysis of the changing competitive landscape in the hotel industry regarding Airbnb. Master's thesis, Dominican University of California.

Lewis, P. (2013) *Couchsurfing: A Sad End to a Great Idea*. See http://thedrawingboard.net/couchsurfing-a-sad-end-to-a-great-idea/ (accessed 18 May 2018).

Li, J., Moreno, A. and Zhang, D. (2015) Agent behavior in the sharing economy: Evidence from Airbnb. Ross School of Business Working Paper Series No. 1298.

Liang, A. (2018) *2 Singapore Airbnb Hosts Plead Guilty to Illegal Rentals*. See https://www.seattletimes.com/business/2-singapore-airbnb-hosts-plead-guilty-to-illegal-rentals/ (accessed 21 May 2018).

Liang, L., Choi, H. and Joppe, M. (2018) Exploring the relationship between satisfaction, trust and switching intention, repurchase intention in the context of Airbnb. *International Journal of Hospitality Management* 69, 41–48.

Liang, S., Schuckert, M., Law, R. and Chen, C. (2017) Be a 'Superhost': The importance of badge systems for peer-to-peer rental accommodations. *Tourism Management* 60, 454–465.

Lim, A. (2018) The potentials of blockchain in the hospitality industry. Unpublished manuscript.

Lines, G. (2015) Hej, Not Hej Da: Regulating Airbnb in the New Age of Arizona Vacation Rentals. *Arizona Law Review* 57, 1163.

Littman, J. (2018) *Hotels and Retailers Hit Hard by California Housing Crisis' Ripple Effect*. See https://www.bisnow.com/san-francisco/news/hotel/san-franciscos-homeless-crisis-is-turning-tourists-away-87174 (accessed 15 May 2018).

Lodder, T. (2002) *Hotels: A Glimpse Behind the Screen*. Leeuwarden: CHN.

Lomas, N. (2018) *Amsterdam to Halve Airbnb-style Tourist Rentals to 30 Nights a Year per Host*. See https://techcrunch.com/2018/01/10/amsterdam-to-halve-airbnb-style-tourist-rentals-to-30-nights-a-year-per-host/ (accessed 1 May 2018).

Loy, T. (2016) *Verbot von Ferienwohnungen in Berlin. Was ist verboten? Was ist erlaubt?* See https://www.tagesspiegel.de/berlin/verbot-von-ferienwohnungen-in-berlin-was-ist-verboten-was-ist-erlaubt/13525762.html (accessed 6 June 2018).

Lubbe Bakker, S. (2016) *Slapend rijk*. See https://www.vpro.nl/programmas/tegenlicht/kijk/afleveringen/2016-2017/slapend-rijk.html (accessed 30 April 2018).

MacCannell, D. (1973) Staged authenticity: Arrangements of social space in tourist settings. *American Journal of Sociology* 79 (3), 589–603.

Mahmud, A.H. (2017) *URA Investigating More Private Residential Properties for Breaking Minimum Stay Law*. See https://www.channelnewsasia.com/news/singapore/airbnb-singapore-investigating-minimum-stay-private-properties-9428320 (accessed 23 May 2018).

Mao, Z. and Lyu, J. (2017) Why travelers use Airbnb again? An integrative approach to understanding travelers' repurchase intention. *International Journal of Contemporary Hospitality Management* 29 (9), 2464–2482.

Martin, C.J. (2016) The sharing economy: A pathway to sustainability or a nightmarish form of neoliberal capitalism? *Ecological Economics* 121, 149–159.

May, K. (2016) *Airbnb and Homeaway Hit Back after Fines in Barcelona*. See https://www.tnooz.com/article/airbnb-and-homeaway-hit-back-after-fines-in-barcelona/ (accessed 8 May 2018).

McAfee, A. and Brynjolfsson, E. (2012) Big data. The management revolution. *Harvard Business Review* 90 (10), 61–67.

McCann, C. (2015) *1979 to 2015 – Average Rent in San Francisco*. See https://medium.com/@mccannatron/1979-to-2015-average-rent-in-san-francisco-33aaea22de0e (accessed 15 May 2018).

McFarland, H. (2017) The Amex decision and the future of antitrust for two-sided platforms. *The Exchange*, Spring, 3–12.

Meek, J. (2014) *Private Island: Why Britain Now Belongs to Someone Else*. London/Brooklyn, NY: Verso Books.

Milikowski, F. and Naafs, S. (2017) *De echte kosten van Pretpark Amsterdam. Wie verdient aan het toerisme?* See https://www.platform-investico.nl/artikel/de-echte-kosten-van-pretpark-amsterdam-2/ (accessed 25 May 2018).

Molz, J. (2013) Social networking technologies and the moral economy of alternative tourism: The case of couchsurfing.org. *Annals of Tourism Research* 43, 210–230.

Moragas, M.d. (2014) *Post-Olympic Barcelona*. Doha: Qatar Tourism Authority/Stenden University Qatar.

Morozov, E. (2014) Don't believe the hype, the 'sharing economy' masks a failing economy. *The Guardian*, 28 September. See https://www.theguardian.com/commentisfree/2014/sep/28/sharing-economy-internet-hype-benefits-overstated-evgeny-morozov (accessed 28 August 2016).

Municipality of Amsterdam (2017) *Overnachtingsbeleid Amsterdam*. See https://www.amsterdam.nl/ondernemen/horeca/horeca/hotelbeleid/ (accessed 4 May 2018).

Municipality of Amsterdam/Airbnb (2014) *Memorandum of Understanding Gemeente Amsterdam & Airbnb*. See https://www.binnenlandsbestuur.nl/Uploads/2016/2/2014-12-airbnb-ireland-amsterdam-mou.pdf (accessed 1 May 2018).

Municipality of Amsterdam/Airbnb (2016) *Overeenkomst Gemeente Amsterdam en Aibnb*. See https://www.amsterdam.nl/publish/pages/593837/overeenkomst_gemeente_amsterdam_en_airbnb.pdf (accessed 1 May 2018).

Nam, T. and Pardo, T. (2011) Conceptualizing Smart City with dimensions of technology, people, and institutions. In *Proceedings of the 12th Annual International Digital Government Research Conference: Digital Government Innovation in Challenging Times* (pp. 282–291). New York: ACM.

NBTC Holland Marketing (2017) *Toerisme in Perspectief*. Den Haag: NBTC.

Neirotti, P., De Marco, A., Cagliano, A.C., Mangano, G. and Scorrano, F. (2014) Current trends in Smart City initiatives: Some stylised facts. *Cities* 38, 25–36.

New York State Attorney General (2014) *Airbnb in the City*. New York: Office of the New York State Attorney General Eric T. Schneiderman.

Nieuwland, S. (2017) Help, Airbnb is taking over the city! A study on the impacts of Airbnb on cities and regulatory approaches. Master's thesis, Utrecht University.

Norton, S. (2015) *Starwood Hotels Using Big Data to Boost Revenue*. See http://blogs.wsj.com/cio/2015/02/10/starwood-hotels-using-big-data-to-boost-revenue/ (accessed 6 April 2016).

Novy, J. (2017) The selling (out) of Berlin and the de- and re-politization of urban tourism in Europe's 'Capital of Cool'. In C. Colomb and J. Novy (eds) *Protest and Resistance in the Tourist City* (pp. 52–72). Abingdon/New York: Routledge.

Novy, J. and Colomb, C. (2017) Urban tourism and its discontents: An introduction. In C. Colomb and J. Novy (eds) *Protest and Resistance in the Tourist City* (pp. 1–30). Abingdon/New York: Routledge.

Nowak, B., Allen, T., Rollo, J. *et al.* (2015) *Global Insight: Who Will Airbnb Hurt More – Hotels or OTAs?* New York: Morgan Stanley Research.

Oates, G. (2016) *How Airbnb is Quietly Positioning Itself to Disrupt the Meetings Industry*. See https://skift.com/2016/02/10/how-airbnb-is-quietly-positioning-itself-to-disrupt-the-meetings-industry/ (accessed 4 March 2018).

O'Connor, P. and Frew, A.J. (2002) The future of hotel electronic distribution: Expert and industry perspectives. *Cornell Hotel and Restaurant Administration Quarterly* 43 (3), 33–45.

OECD (2009) *Policy Roundtables: Two-sided Markets. DAF/COMP (2009) 20*. Paris: OECD.

Olson, K. (2012) *National Study Quantifies the 'Sharing Economy' Movement*. See https://www.prnewswire.com/news-releases/national-study-quantifies-the-sharing-economy-movement-138949069.html (accessed 24 February 2018).

O'Neill, J. and Ouyang, Y. (2016) *From Air Mattresses to Unregulated Business: An Analysis of the Other Side of Airbnb*. University Park, PA: Pennsylvania State University.

Oskam, J. and Boswijk, A. (2016) Airbnb: The future of networked hospitality businesses. *Journal of Tourism Futures* 2 (1), 22–42.

Oskam, J. and Zandberg, T. (2016) Who will sell your rooms? Hotel distribution scenarios. *Journal of Vacation Marketing* 22 (3), 265–278.

Oskam, J., Van der Rest, J. and Telkamp, B. (2018) What's mine is yours – but at what price? Dynamic pricing behavior as an indicator of Airbnb host professionalization. *Journal of Revenue & Pricing Management* 17 (5), 311–328.

Pan, B. (2018) *Tourism to the US is in a 'Trump Slump' – Truth or Fiction?* See http://theconversation.com/tourism-to-the-us-is-in-a-trump-slump-truth-or-fiction-92254 (accessed 10 September 2018).

Parool (2018) *Het Parool lanceert Kieswijzer voor Amsterdam*. See https://www.parool.nl/verkiezingen/het-parool-lanceert-kieswijzer-voor-amsterdam~a4573012/ (accessed 2 May 2018).

Passiak, D. (2017) *Belong Anywhere – The Vision and Story Behind Airbnb's Global Community*. See https://medium.com/cocreatethefuture/belong-anywhere-the-vision-and-story-behind-airbnbs-global-community-123d32218d6a (accessed 2 February 2017).

Peltier, D. (2015) *3 Airbnb Hosts on What Guests Demand Now*. See https://skift.com/2015/11/12/3-airbnb-hosts-on-what-guests-want-now/ (accessed 13 November 2015).

Perloth, N. (2011) *Non-Profit CouchSurfing Raises Millions In Funding*. See https://www.forbes.com/sites/nicoleperlroth/2011/08/24/non-profit-couchsurfing-raises-millions-in-funding/#23a8ce413e3d (accessed 18 May 2018).

Peterson, R. (2005) In search of authenticity. *Journal of Management Studies* 42 (5), 1083–1098.

Pezenka, I., Winkler, P. and Weismayer, C. (2017) *Erlebnis als Versprechen der Sharing Economy – Eine Big Five Studie am Beispiel von Airbnb NutzerInnen*. Salzburg: Forschungsforum der Österreichischen Fachhochschulen.

Picascia, S., Romano, A. and Teobaldi, M. (2017) The airification of cities: Making sense of the impact of peer to peer short term letting on urban functions and economy. In *Proceedings of the Annual Congress of the Association of European Schools of Planning*, Lisbon, 11–14 July.

Pinkster, F.M. and Boterman, W.R. (2017) When the spell is broken: Gentrification, urban tourism and privileged discontent in the Amsterdam canal district. *Cultural Geographies* 24 (3), 457–472.

Poon, L. (2017) *Singapore, City of Sensors*. See https://www.citylab.com/life/2017/04/singapore-city-of-sensors/523392/ (accessed 26 May 2018).

Porter, M.E. (2001) Strategy and the internet. *Harvard Business Review* 79 (3), 62–78.

Port of Amsterdam (2016) *Full Steam Ahead. Port of Amsterdam Annual Report 2016*. Amsterdam: Port of Amsterdam.

PWC (2018) *Best Placed to Grow? European Cities Hotel Forecast for 2018 and 2019*. London: PriceWaterhouseCoopers.

Quattrone, G., Proserpio, D., Quercia, D., Capra, L. and Musolesi, M. (2016) *Who Benefits from the Sharing Economy of Airbnb?* (pp. 1385–1394). In *Proceedings of the 26th International ACM Conference on World Wide Web (WWW) 2016*, Montreal.

Realty Shares (2016) *The Rise of the Professional Airbnb Investor*. See https://www.realtyshares.com/blog/the-rise-of-the-professional-airbnb-investor/ (accessed 12 February 2016).

Redeker Sellner Dahs (2017) *Kehrtwende beim Berliner Zweckentfremdungsverbot: Home Sharern wird Vermietung der Hauptwohnung an insgesamt 182 Tagen pro Jahr gestattet*. See https://www.redeker.de/main-V2.php/de/news/pm20170908.html (accessed 8 June 2018).

ResonanceCo (2017) *Future of U.S. Millenial Travel: A Survey of America's Fastest Growing Tourism Demographic*. Vancouver: ResonanceCo.

Richard, B. and Cleveland, S. (2016) The future of hotel chains: Branded marketplaces driven by the sharing economy. *Journal of Vacation Marketing* 22 (3), 239–248.

Rifkin, J. (2000) *The Age of Access: The New Culture of Hypercapitalism*. New York: Penguin Putnam.

Rijk, M.d. and Venema, A.-J. (2016) *De kosten in de zorg*. See https://www.groene.nl/artikel/de-kosten-in-de-zorg (accessed 23 June 2018).

Rochet, J. and Tirole, J. (2003) Platform competition in two-sided markets. *Journal of the European Economic Association* 1 (4), 990–1029.

Rochet, J.-C. and Tirole, J. (2004) Two-sided markets: An overview. IDEI Working Paper No. 258.

Rochet, J.-C. and Tirole, J. (2006) Two-sided markets: A progress report. *RAND Journal of Economics* 37 (3), 645–667.

Romero, V. (2017) *Golpe a las plataformas con pisos ilegales: Una sentencia obliga a HomeAway a retirarlos*. See https://www.elconfidencial.com/espana/comunidad-valenciana/2017-12-21/someway-airbnb-apartamentos-ilegales-golpe-sentencia-valencia_1496569/ (accessed 2 June 2018).

Rosen, K.T., Sakamoto, R. and Bank, D. (2013) *Short-Term Rentals and Impact on the Apartment Market*. Berkeley, CA: Rosen Consulting Group.

Roth, H. and Fishbin, M. (2015) *Global Hospitality Insights. Top Thoughts for 2015*. New York: EY Global Real Estate, Hospitality & Construction Center.

Said, C. (2015) *Prop. F: S.F. Voters RejectMmeasure to Restrict Airbnb Rentals*. See https://www.sfgate.com/bayarea/article/Prop-F-Measure-to-restrict-Airbnb-rentals-6609176.php (accessed 15 May 2018).

Said, C. (2018a) *A Leaner Vacation-rental Market*. See https://www.sfchronicle.com/business/article/SF-short-term-rentals-transformed-as-Airbnb-12617798.php (accessed 18 May 2018).

Said, C. (2018b) *Airbnb Loses Thousands of Hosts in SF as Registration Rules Kick In*. See https://www.sfchronicle.com/business/article/Airbnb-loses-thousands-of-hosts-in-SF-as-12496624.php (accessed 15 May 2018).

Sala i Martin, X. (1998) *Independència de Catalunya: La Viabilitat Econòmica. Conference at Omnium Cultural, February 1998*. See http://www.columbia.edu/~xs23/papers/independ.htm (accessed 17 August 2013).

Sala i Martin, X. (2001) *Catalanisme Obert al Segle XXI: L'Economia'. Paper Read at the Conference 'Catalanisme Obert al Segle XXI, organitzada per la Fundació Catalunya Oberta, Girona, 23 de Novembre del 2001*. See http://www.columbia.edu/~xs23/catala/articles/2001/FCO/Fundacio%20Catalunya%20Oberta%20Paper.pdf (accessed 18 August 2013).

Samaan, R. (2015) *Airbnb, Rising Rent and the Housing Crisis in Los Angeles*. Los Angeles, CA: LAANE.

Sampson, H. (2017) *More Companies Are Allowing Travelers to Use Sharing Economy Services*. See https://skift.com/2017/02/01/more-companies-are-allowing-travelers-to-use-sharing-economy-services/ (accessed 4 March 2018).

Satama, S. (2014) *Consumer Adoption of Access-based Consumption Services – Case AirBnB*. Espoo: Aalto University.

Schaal, D., Stone, R. and Ting, D. (2018) *Skift Presents: The Airbnb Threat. Who Will Ultimately Win the Online Travel Battle?* See https://3rxg9qea18zhtl6s2u8jammft-wpengine.netdna-ssl.com/wp-content/uploads/2018/03/skiftCall-airbnb.pdf (accessed 23 March 2018).

Scheiber, N. (2017) How Uber uses psychological tricks to push its drivers' buttons. *New York Times*, 2 April. See https://www.nytimes.com/interactive/2017/04/02/technology/uber-drivers-psychological-tricks.html (accessed 19 April 2017).

Schnitzler, H. (2015) *Het Digitale Proletariaat*. Amsterdam: De Bezige Bij.

Schumpeter, J. (1942) *Capitalism, Socialism and Democracy*. New York/London: Harper.

Servihabitat (2017) *Mercado de alquiler residencial en España*. Barcelona/Madrid: Servihabitat.

SF Residential Rent Stabilization and Arbitration Board (2017) *Twenty Years of Rent Board Annual Reports on Eviction Notices*. San Francisco, CA: San Francisco Residential Rent Stabilization and Arbitration Board.

ShareNL (2015) *Amsterdam Sharing City Projects*. See http://www.sharenl.nl/amsterdam-sharing-city-projects/ (accessed 30 April 2018).

Shaw, L. (2017) *Music Industry Soars into Year 3 of Recovery Thanks to Spotify*. See https://www.bloomberg.com/news/articles/2017-09-20/music-industry-soars-into-year-3-of-recovery-thanks-to-spotify (accessed 16 January 2018).

Shaw, R. (2014) *Mayor, Supervisor Kim Give Teeth to Airbnb Law*. See http://www.beyondchron.org/mayor-lee-supe-kim-win-enforceable-airbnb-law/ (accessed 1 February 2018).

Sheivachman, A. (2017) *Airbnb Business Travel Transactions Doubled Since 2014*. See https://skift.com/2017/01/27/airbnb-business-travel-transactions-doubled-since-2014/ (accessed 27 January 2017).

Singapore Legal Advice (2017) *Is Airbnb Illegal in Singapore?* See https://singaporelegaladvice.com/law-articles/is-airbnb-illegal-singapore (accessed 23 May 2018).

Singapore Smart Nation (2018) *Smart Nation Singapore. Many Smart Ideas, One Smart Nation*. See https://www.smartnation.sg/ (accessed 28 April 2018).

Singapore Tourism Board (2018) *Annual Report on Tourism Staristics 2016*. Singapore: Singapore Tourism Board.

Slee, T. (2015) *What's Yours is Mine. Against the Sharing Economy*. New York/London: OR Books.

Smith, A.J. (2018) *The Next Trend In Travel Is … Don't*. See https://brightthemag.com/the-next-trend-in-travel-is-dont-226d4aba17f6 (accessed 13 September 2018).

Smith, H. (2017) Santorini's popularity soars but locals say it has hit saturation point. *The Guardian*, 28 August. See https://www.theguardian.com/world/2017/aug/28/santorini-popularity-soars-but-locals-say-it-has-hit-saturation-point (accessed 29 August 2017).

Smith, N. (2005) *The New Urban Frontier: Gentrification and the Revanchist City*. London: Routledge.

Smith, N. (2018) *Econ 101 No Longer Explains the Job Market*. See https://www.bloomberg.com/view/articles/2018-04-05/supply-and-demand-does-a-poor-job-of-explaining-depressed-wages (accessed 20 June 2018).

Sondermeijer, V. (2017) *Airbnb: In 2016 1,4 miljoen Nederlandse boekingen*. See https://www.nrc.nl/nieuws/2017/02/21/airbnb-14-miljoen-nederlandse-boekingen-in-2016-a1546972 (accessed 22 May 2017).

Soriano, D. (2015) *Barcelona, a la caza del piso turístico: inspectores puerta a puerta y multas de hasta 90.000 euros*. See https://www.libremercado.com/2015-05-15/barcelona-sale-a-la-caza-del-piso-turistico-con-multas-de-hasta-90000-euros-1276548042/ (accessed 18 October 2015).

Sottek, T. (2014) *Uber Has an Army of at Least 161 Lobbyists and They're Crushing Regulators*. See http://www.theverge.com/2014/12/14/7390395/uber-lobbying-steamroller (accessed 19 October 2015).

Spotswood, B. (2017) *2017 San Francisco 'Homeless Census' Reveals That Despite Numbers, Things Are Worse, Not Better*. See http://sfist.com/2017/06/26/2017_san_francisco_homeless_census.php (accessed 15 May 2018).

Sritama, S. and Katharansiporn, K. (2018) Whither Airbnb? *Bangkok Post*, 28 May, p. B1.

Stern, J. (2010) *AirBnb Benefits from Social Proof Theory*. See http://josephstern.com/wp-content/uploads/2010/10/Social-Proof-Theory-Paper.pdf (accessed 24 August 2015).

Stors, N. and Kagermeier, A. (2015) Motives for using Airbnb in metropolitan tourism – why do people sleep in the bed of a stranger? *Regions Magazine* 299 (1), 17–19.

Strong, C. (2014) *Airbnb and Hotels: What to Do about the Sharing Economy*. See https://www.wired.com/insights/2014/11/hotels-sharing-economy/ (accessed 21 March 2018).

Sundararajan, A. (2014) Trusting the 'sharing economy' to regulate itself. *New York Times*, 3 March. See https://economix.blogs.nytimes.com/2014/03/03/trusting-the-sharing-economy-to-regulate-itself/ (accessed 3 June 2018).

Sundararajan, A. (2016) *The Sharing Economy: The End of Employment and the Rise of Crowd-Based Capitalism*. Cambridge, MA: MIT Press.

Sunderland, J. (2014) *Shattered Dreams: Impact of Spain's Housing Crisis on Vulnerable Groups*. See https://www.hrw.org/report/2014/05/27/shattered-dreams/impact-spains-housing-crisis-vulnerable-groups (accessed 23 June 2018).

Tam, D. (2014) *SF: Airbnb, Craigslist Rentals Need to 'Play by the Rules' (Q&A)*. See https://www.cnet.com/news/airbnb-craigslist-need-to-play-by-the-rules-in-san-francisco-q-a/ (accessed 14 May 2018).

Taylor, J. (2001) Authenticity and sincerity in tourism. *Annals of Tourism Research* 28 (1), 7–26.

Teubner, T., Hawlitschek, F. and Dann, D. (2017) Price determinants on Airbnb: How reputation pays off in the sharing economy. *Journal of Self-Governance and Management Economics* 5 (4), 53–80.

The Economist (2016) American house prices: Realty check. *The Economist*, 24 August. See https://www.economist.com/blogs/graphicdetail/2016/08/daily-chart-20 (accessed 15 May 2018).

The Economist (2018) Big Mac Index. *The Economist*, 17 January. See https://www.economist.com/content/big-mac-index (accessed 23 May 2018).

Tim, H.C. and Sussman, S. (2014) *What Makes an Asian Tiger? Singapore's Unlikely Economic Success Lies in its History*. See https://www.forbes.com/sites/forbesasia/2014/07/10/what-makes-an-asian-tiger-singapores-unlikely-economic-success-lies-in-its-history/#4e38a2f76697 (accessed 23 May 2018).

Ting, D. (2016) *Airbnb Tests Hotel-Style Packaging and Amenities in Sonoma, CA*. See https://skift.com/2016/04/27/airbnb-tests-hotel-style-packaging-and-amenities-in-sonoma-ca/ (accessed 27 March 2018).

Ting, D. (2017a) *Airbnb Is Becoming an Even Bigger Threat to Hotels Says a New Report*. See https://skift.com/2017/01/04/airbnb-is-becoming-an-even-bigger-threat-to-hotels-says-a-new-report/ (accessed 25 January 2017).

Ting, D. (2017b) *Hyatt Is Investing in the Sharing Economy Again, This Time With Oasis*. See https://skift.com/2017/08/03/hyatt-is-investing-in-the-sharing-economy-again-this-time-with-oasis/ (accessed 26 March 2018).

Ting, D. (2017c) *Airbnb Acquires Vacation Rental Company Luxury Retreats, Officially Moves into Luxury*. See https://skift.com/2017/02/16/airbnb-acquires-vacation-rental-company-luxury-retreats-officially-moves-into-luxury/ (accessed 28 March 2018).

Ting, D. (2017d) *Short-Term Rental Ban in Berlin for Airbnb and Others Appears to Be Eroding*. See https://skift.com/2017/09/11/short-term-rental-ban-in-berlin-for-airbnb-and-others-appears-to-be-eroding/ (accessed 11 September 2017).

Ting, D. (2018a) *Airbnb Is Set to Launch a New Tier of Select Properties*. See https://skift.com/2018/02/13/airbnb-is-set-to-launch-a-new-tier-of-select-properties/ (accessed 25 March 2018).

Ting, D. (2018b) *Airbnb Plus' Risky Bet to Push Homesharing to a New Level*. See https://skift.com/2018/04/02/airbnb-plus-risky-bet-to-push-homesharing-to-a-new-level/ (accessed 2 April 2018).

Ting, D. (2018c) *This Is How Airbnb Plans to Win Over Hotels*. See https://skift.com/2018/02/27/this-is-how-airbnb-plans-to-win-over-hotels/ (accessed 23 March 2018).

Trading Economics (2018) *Singapore GDP Growth Rate*. See https://tradingeconomics.com/singapore/gdp-growth (accessed 23 May 2018).

Trulia (2018) *San Francisco Real Estate Market Overview*. See https://www.trulia.com/real_estate/San_Francisco-California/ (accessed 15 May 2018).

Tucker, C. (2018) *Why Network Effects Matter Less Than They Used To*. See https://hbr.org/product/why-network-effects-matter-less-than-they-used-to/H04EUV-PDF-ENG (accessed 27 June 2018).

Turnbull, D. and Heinze, M. (2015) *Hospitality Analytics: How Data Can Make Hotels Smarter*. New York: Skift & Snapshot.

Tussyadiah, I. (2015) An exploratory study on drivers and deterrents of collaborative consumption in travel. In B. Stangl and J. Pesonen (eds) *Information and Communication Technologies in Tourism* (pp. 817–830). Cham: Springer.

Tussyadiah, I. and Pesonen, J. (2016) Drivers and barriers of peer-to-peer accommodation stay – an exploratory study with American and Finnish travellers. *Current Issues in Tourism* 21 (6), 703–720.

UNWTO (2011) *Tourism Towards 2030. Global Overview*. See http://media.unwto.org/sites/all/files/pdf/unwto_2030_ga_2011_korea.pdf (accessed 4 April 2018).

UNWTO (2018) *2017 International Tourism Results: The Highest in Seven Years*. See http://media.unwto.org/press-release/2018-01-15/2017-international-tourism-results-highest-seven-years (accessed 2 May 2018).

URA (2018) *Consultation Exercise on Proposed Regulatory Framework for the Use of Private Residential Properties as Short-Term Accommodation*. Singapore: Urban Redevelopment Authority.

US Servas (2018) *General FAQ*. See https://usservas.org/faq/11 (accessed 18 May 2018).

Van Burgeler, M. (2017) A study on the use of storytelling in Airbnb's marketing campaigns. BA dissertation, Hotelschool The Hague.

Van den Broek, V. (2017) Airbnb user satisfaction in Amsterdam. BA dissertation, Hotelschool The Hague.

Van den Bent, C. (2016) Motives for Airbnb users in Amsterdam to choose Airbnb. BA dissertation, Hotelschool The Hague.

Van der Bijl, V. (2016) The effect of Airbnb on house prices in Amsterdam: A study of the side effects of a disruptive start-up in the new sharing economy. Master's thesis, University of Amsterdam.

Van der Heijden, K. (2005) *Scenarios: The Art of Strategic Conversation*. New York: John Wiley.

Van Loenen, M. (2016) Airbnb in Amsterdam: Visitor profiles. BA dissertation, Hotelschool The Hague.

Vekshin, A. (2015) *Airbnb's $8 Million Plus Fight Over Proposed Rules in San Francisco*. See http://skift.com/2015/10/23/airbnbs-8-million-plus-fight-over-proposed-rules-in-san-francisco/ (accessed 23 October 2015).

Verdú, D. (2015) El 'efecto Airbnb' en el vecindario. *El País*, 10 December. See https://elpais.com/economia/2015/12/10/actualidad/1449738303_311413.html (accessed 15 December 2015).

Verwaltungsgericht Berlin (2016) *Zweckentfremdung: Anspruch auf Erteilung von Ausnahmegenehmigungen für Ferienwohnungszwecke bei Zweitwohnungen* (No. 34/2016). See https://www.berlin.de/gerichte/verwaltungsgericht/presse/pressemitteilungen/2016/pressemitteilung.507033.php (accessed 8 June 2018).

Visjø, C.T. and Slevigen, H.H. (2017) *How to Create Loyalty in the Sharing Economy? A Study of Emotions, Satisfaction and Commitment among Airbnb Customers*. Oslo: Norwegian Business School.

Visser, B. (2017) Airbnb. BA dissertation, Hotelschool The Hague.

Vranken, J. and Van der Most, K. (2016) Interview, 24 March 2016.

Wachsmuth, D., Chaney, D., Kerrigan, D., Shillolo, A. and Basalaev-Binder, R. (2018) *The High Cost of Short-Term Rentals in New York City*. Montreal: McGill University.

Wallsten, S. (2015) *The Competitive Effects of the Sharing Economy: How is Uber Changing Taxis?* See https://www.researchgate.net/publication/279514652_The_Competitive_Effects_of_the_Sharing_Economy_How_is_Uber_Changing_Taxis (accessed 4 September 2015).

Wang, D. and Nicolau, J. (2017) Price determinants of sharing economy based accommodation rental: A study of listings from 33 cities on Airbnb.com. *International Journal of Hospitality Management* 62, 120–131.

Wang, N. (1999) Rethinking authenticity in tourism experience. *Annals of Tourism Research* 26 (2), 349–370.

Ward, J. (2016) *U.K. Ponders Big Data to Measure the Economy of Uber and Airbnb*. See http://www.bloomberg.com/news/articles/2016-04-05/u-k-ponders-big-data-to-measure-the-economy-of-uber-and-airbnb (accessed 6 April 2016).

Welch, L. (2014) *Entrepreneur Designs Upscale Hotels for Budget Travelers*. See https://www.inc.com/magazine201406/liz-welch/citizenm-low-cost-high-end-hotels.html (accessed 24 March 2018).

Wen-Yi, L. and Wei, L.D. (2018) *Mixed Reactions to URA Proposal to Allow Short-term Condominium Rentals*. See https://www.straitstimes.com/singapore/mixed-reactions-to-ura-proposal-to-allow-short-term-condominium-rentals (accessed 23 May 2018).

Whyte, P. (2016) *Hyatt Invested $15 Million in Cash-Burning Onefinestay Before Accor Acquisition*. See https://skift.com/2016/12/02/hyatt-invested-15-million-in-cash-burning-onefinestay-before-accor-acquisition/ (accessed 24 March 2018).

Whyte, P. (2018) *Video: Airbnb's European Boss on the Clash with Online Travel Giants*. See https://skift.com/2018/06/11/video-airbnbs-european-boss-on-the-clash-with-online-travel-giants/ (accessed 11 June 2018).

World Bank Group (2017) *Doing Business: Economy Rankings*. See http://www.doing-business.org/rankings (accessed 23 May 2018).

World Bank Group (2018) *GDP per capita (current US$)*. See https://data.worldbank.org/indicator/NY.GDP.PCAP.CD (accessed 23 May 2018).

Xiang, Y. and Dolnicar, S. (2017) Networks in China. In S. Dolnicar (ed.) *Peer-to-Peer Accommodation Networks: Pushing the Boundaries* (pp. 149–159). Oxford: Goodfellow.

Xiang, Z., Schwartz, Z., Gerdes, J.H. and Uysal, M. (2015) What can big data and text analytics tell us about hotel guest experience and satisfaction? *International Journal of Hospitality Management* 44, 120–130.

Xie, K. and Mao, Z. (2017) The impacts of quality and quantity attributes of Airbnb hosts on listing performance. *International Journal of Contemporary Hospitality Management* 29 (9), 2240–2260.

Yang, S. and Ahn, S. (2016) Impact of motivation for participation in the sharing economy and perceived security on attitude and loyalty toward Airbnb. *International Information Institute (Tokyo), Information Koganei* 19 (12), 5745–5750.

Yannopoulou, N., Moufahim, M. and Bian, X. (2013) User-generated brands and social media: Couchsurfing and AirBnB. *Contemporary Management Research* 9 (1).

Yeoman, I., Brass, D. and McMahon-Beattie, U. (2007) Current issue in tourism: The authentic tourist. *Tourism Management* 28 (4), 1128–1138.

Yuan, L. (2017) *The Man who Invented the Boutique Hotel has a New Concept of 'Luxury for All' for the Digital Age*. See https://qz.com/1002728/the-man-who-invented-the-boutique-hotel-has-a-new-concept-of-luxury-for-all-for-the-digital-age/ (accessed 24 March 2018).

Zaleski, O. (2017a) *Airbnb-Branded Apartment Buildings Are Coming to the U.S.* See https://www.bloomberg.com/news/articles/2017-10-12/airbnb-branded-apartment-buildings-are-coming-to-the-u-s (accessed 10 December 2017).

Zaleski, O. (2017b) *Airbnb Readies a Premium Tier to Compete More With Hotels, Sources Say.* See https://www.bloomberg.com/news/articles/2017-06-21/airbnb-said-to-ready-a-premium-tier-to-compete-more-with-hotels (accessed 25 March 2018).

Zervas, G., Proserpio, D. and Byers, J. W. (2015) *A First Look at Online Reputation on Airbnb, Where Every Stay is Above Average.* See http://collaborativeeconomy.com/wp/wp-content/uploads/2015/04/Byers-D.-Proserpio-D.-Zervas-G.2015.A-First-Look-at-Online-Reputation-on-Airbnb-Where-Every-Stay-is-Above-Average.Boston-University.pdf (accessed 24 August 2015).

Zervas, G., Proserpio, D. and Byers, J. (2017) The rise of the sharing economy: Estimating the impact of Airbnb on the hotel industry. *Journal of Marketing Research* 54 (5), 687–705.

Zubizarreta, I., Seravalli, A. and Arrizabalaga, S. (2015) Smart City concept: What it is and what it should be. *Journal of Urban Planning and Development* 142 (1), 04015005.

Zukin, S. (2008) Consuming authenticity, from outposts of difference to means of exclusion. *Cultural Studies* 22 (5), 724–748.

Zukin, S. (2009) New retail capital and neighbourhood change: Boutiques and gentrification in New York City. *City and Community* 8 (1), 47–64.

Index

60 Days, 60
9Flats, 26

Abad Liñán, José Manuel, 118, 151
accessibility, 96
Accor, 24, 26, 73, 170
Ahn, Sungsook, 47, 170
Airbnb for Business, 48, 49
Airbnb Plus, 59, 74, 75, 155, 169
Airbnb Select, 74
airbnbhell.com, 48
Airbnbutler, 60
AirDNA, 2, 13–15, 56–58, 65, 68,
    81–85, 87, 90, 101, 102, 108, 109,
    119, 137, 152
Alibaba, 20
Allor, Brett, 128, 152
Amazon, 27, 42, 45, 46, 61
Amazon Mechanical Turk, 42, 45, 46, 61
American Hotel and Lodging
    Association, 67, 70
Amex, 23, 164
Amsterdam, 1, 2, 12–18, 43, 49–51, 53,
    54, 57, 58, 60, 65, 68–70, 72, 75,
    78, 80–82, 84–88, 90, 94, 97, 98,
    100, 101, 105–113, 115, 116, 118,
    119, 122, 124, 125, 132, 144, 145,
    148, 151–155, 157, 159, 162–165,
    167, 169
Amsterdam in Progress, 110, 152
Amsterdam Marketing, 2, 152
Anaheim, 97, 153
Anderson, Chris, 76, 152
Angelidou, Margarita, 138, 152
Angkor Wat, 148
Apple, 22, 27, 33, 142
Araya, Daniel, 10, 18, 152
Arias-Sans, Albert, 14, 64, 80, 116, 152
Armstrong, Mark, 21, 152
Arran, 114, 117, 152

Associated Press, 137, 152
AT5, 98, 153
Aunión, J.A., 91, 153
Austin, 26, 55, 58, 69
authenticity, 1, 24, 31–34, 36, 37, 39–41,
    44, 45, 48, 53, 54, 71, 72, 85, 111, 112,
    130, 141, 142, 145, 147, 150, 159, 160,
    162, 163, 165, 168, 170, 171
Au-Yong, Rachel, 137, 153
AvalonBay, 60
Awaykey, 60
Azores, 38, 40

Badia, Evylen, 41, 153
Bakker, Dirk, 2, 12–15, 25, 55–58, 63–65,
    68–70, 81–85, 87, 99–102, 108, 109,
    119, 153
Balanzó, Rafael de, 116, 153
Bali, 135
Baquero, Camilo S., 116, 153
Barcelona, 6, 16, 19, 46, 52, 55, 68, 80,
    83, 86–88, 103, 114–121, 124, 125,
    132, 144, 145, 148, 151–153, 158,
    160, 161, 163, 164, 167
Barcelona Activa, 114, 153
Barcelona en Comú, 116, 153
Bardhi, Fleura, 9, 153, 158
Bari, 81
Barragan, Bianca, 97, 153
Barron, Kyle, 90, 153
Bauwens, Michel, 10, 11, 18, 152, 154, 162
Beck, Luisa, 101, 154
Bee Token, 29, 30, 130, 154
BeeNest, 29, 30, 130, 154
Beer, Tim de, 54, 154
Beijing, 26, 135
Belgium, 146
Belk, Russell, 8, 9, 154
BeMate, 73, 154
Benner, Katie, 67, 73, 154

Berg, Janine, 61, 63, 154
Bergman, Justin, 25, 154
Berlin, 13–15, 26, 57, 58, 65, 68, 69, 75, 80, 82, 84–87, 97, 100–102, 153, 154, 159, 161–164, 166, 168, 169
BeWelcome, 129, 154
Beyond by Airbnb, 74
billboard effect, 76, 152
Billock, Jennifer, 129, 154
BiLove, 146
Bishop, Peter, 5, 28, 154, 161
Bitcoin, 29, 62
Blickhan, Micki, 85, 154
blockchain, 28–30, 130, 161, 163
Bloomberg, 102, 154, 167, 170, 171
BMG, 27
BMW, 33
Bnbmanager, 60
Bolivia, 145
BookCrossing, 10
Booking, 23–27, 74–77, 112
Booth, Kwan, 18, 123, 155
Bosa, Deirdre, 67, 155
Bostoen, Friso, 23, 155
Boston, 44, 151
Boswijk, Albert, 11, 12, 19, 91, 111, 120, 165
Boterman, Willem R., 112, 165
Botsman, Rachel, 7, 9, 10, 11, 19, 42, 43, 155
Bouma, Floor, 98, 155
Bournemouth, 57
Bowers, Brittany, 91, 155
Bowles, Nellie, 127, 155
Brinklow, Adam, 128, 155
British Hospitality Association, 70
Bronx, 40
Brooklyn, 40, 89, 90
Brousseau, Fred, 99, 123, 155
Brugnaro, Luigi, 110
Brunton, John, 110, 155
Brussels, 68, 70, 87, 155
Brynjolfsson, Erik, 16, 164
Budapest, 64, 158
Buhalis, Dimitrios, 23, 155
Bulgaria, 122
Bureau of Labour Statistics, 125
Business Travel IQ, 70, 155

Caillaud, Bernard, 20, 155
Calder, Simon, 74, 155

California Federation of Teachers, 6, 123
California Nurses Association, 123
Camden Property Trust, 60
Carey, Meredith, 74, 98, 104, 155, 156
Carroll, Bill, 23, 156
Carson, Biz, 60, 156
Castillo, Ignacio A., 91, 156
Catà, Josep, 118, 156
Cave, Bryan, 124, 156
CEIBS, 26, 156
Certify, 49
Chadha, Rattan, 71
Chai Jia Jih, 135
Channel News Asia, 98, 137, 156
Chernyshenko, Olexander S., 134, 156
Chesky, Brian, 75, 93, 96
Chiarelli, Nick, 45, 156
Chicago, 58, 69
China, 25, 26, 102, 134–136, 154, 157, 170
Chiu, David, 123, 154
Christides, Giorgos, 90, 156
Cinque Terre, 88
CitizenM, 71, 72, 170
City Attorney of San Francisco, 103, 124, 156
Ciutat per a qui l'habita, 79, 156
Clampet, Jason, 52, 62
Clark, Patrick, 52, 156
Clemente, Yolanda, 91, 153
Cleveland, Shane, 72, 166
Coca, Nithin, 122, 129, 156
Cócola Gant, Agustín, 115, 116, 157
Cohen, Leonard, 96, 157
Colliers International, 2, 68, 153
Colomb, Claire, 86, 160, 164
commissions, 5, 18, 25, 27, 59, 74, 77, 119, 129–131
commodification, 9, 92, 144–149, 160
compression nights, 70
Conley, Chip, 73
Consigli, Matt, 24, 48, 157
consumer protection, 17, 22, 48, 93, 96, 104, 144,
convergence, 74, 77, 78, 112, 113, 132, 140, 143, 144, 149
Cooperativa Integral Catalana, 11
Copenhagen, 87
Corbett, Sara, 54, 157
Cornell University, 76, 152, 156, 165

Couchsurfing, 10, 11, 122, 129, 130, 154, 156, 157, 162, 163, 164, 165, 170
Cox, Murray, 1, 14, 54, 65, 90, 97, 100, 101, 103, 157
Craigslist, 62, 168
cruise tourism, 79, 86–88, 109, 110, 112, 114, 115, 120
cults, 31, 33, 34, 36, 38, 40, 41, 67, 130, 142, 153
Custer, C., 25, 157
Cyber Report, 97, 157

Daniels, Jeff, 139, 157
Dator, Jim, 5, 144, 157
Davies, S., 17, 157
day visitors, 50, 86, 112, 128
Dekker, Daphne Maria, 72, 157
Delgado-Medrano, Héber Manuel, 89, 153
Deloitte, 45, 62, 157, 163
Demurtas, Alice, 91, 157
Denmark, 134
Denyer, Simon, 139, 157
deregulation, 16, 17, 93
Diário da República, 98, 158
Dillahunt, Tawanna R., 62, 158
Dimanche, Frédéric, 71, 159
disintermediation, 23, 28, 29, 75, 160
Disney, 32, 33
Dobbins, James, 54, 158
Dolnicar, Sara, 8, 26, 30, 64, 158, 160, 170
Donato, James, 124, 158
Dredge, Dianne, 8, 160
Dublin, 87
Dudás, Gábor, 64, 158
Duran, Pere, 115, 158

Ebay, 11, 27
Eckhardt, Giana M., 9, 153, 158
Economist, The, 125, 126, 136, 155, 168
Economista, El, 118, 158
Edelman, Benjamin, 65, 94–96, 158
Edwards, Jim, 89, 158
Eisenmann, Thomas, 20–23, 27, 158
EMI, 27
Emprendedores, 71, 158
enforcement, 16, 88, 94, 96, 97, 99, 100, 107, 120, 123, 124, 136, 137, 139, 143, 160, 163, 167
envelopment, 23–26, 30

Ert, Eyal, 48, 158
Espinosa, Tristan P., 95, 158
Etsy, 11, 61
Evans, David, 20, 22, 27, 158, 159
Expedia, 23, 24, 26, 74, 75, 77
extraction, 10, 18, 27
EY, 44, 45, 80, 159, 166

Facebook, 10, 11, 20, 27, 46
Fanning, Shawn, 27
Fedorova, Tanja, 109, 159
Finland, 45
Fishbin, Michael, 45, 166
Florence, 80, 81
Forgacs, Gábor, 71, 159
Four Futures Archetypes, 5
Fradkin, Audrey, 48, 159
Framework Foresight Method, 5
France, 129, 130
France 24, 17, 159
Frankel, Matthew, 131, 159
Frenken, Koen, 146, 159
Frew, Andrew J., 23, 165
Füller, Henning, 85, 159
Future Foundation, 45

Gansky, Lisa, 9, 43, 159
Garcés, Eliana, 16, 23, 159
Garfield, Leanna, 71, 159
Garrett-Price, Kwame, 54, 159
Gebbia, Joe, 7, 11, 159
Genoa, 81, 87
gentrification, 80, 84–86, 89, 90, 92, 111, 115, 116, 120, 121, 124, 125, 128, 152, 157, 159, 161, 163, 165, 167, 171
Geradin, Damien, 95, 96, 158
Gibbs, Chris, 58, 64, 159
Gier, Matthijs de, 54, 154
Gilheany, John, 65, 159
Gilmore, James H., 33, 39, 159
Glaeser, Edward L., 124, 159
Glass, Ruth, 85, 159
Global Business Network approach, 5
Global Business Travel Association, 49
Gnutella, 28
Goffman, Erving, 32, 159
Goldman, Alexandra, 127, 128, 155
Goodwin, Tom, 20, 160
Google, 24, 27, 128, 159
Grand Ferdinand, 76

Gravari-Barbas, Maria, 86, 160
Green, Matthew, 27, 160
Griswold, Alison, 102, 103, 160
Groen, Willem Pieter de, 61, 157
Grokster, 95
guest satisfaction, 16, 47, 48, 51, 163, 169, 170
Guest to Guest, 129, 130, 132, 160
Gunter, Ulrich, 25, 47, 59, 72, 160
Gutiérrez, Javier, 80, 83, 160
Guttentag, Daniel, 2, 17, 24, 46, 47, 49, 51, 71, 72, 132, 159, 160
Gyimóthy, Szilvia, 8, 160
Gyourko, Joseph, 124, 159

Hajibaba, Homa, 30, 160
Hall, Colin Michael, 33, 160
Hamari, Juho, 45, 160
Hamburg, 26
Hantman, D., 123, 160
Harboe Sorensen, Vincent, 60, 160
Harley Davidon, 142
Hartung, Richard, 133, 135, 160
Harvey, David, 145, 147, 160
Hawaii, 39, 40
Haywood, Jessica, 70, 160
Heckman, Matt, 21, 161
Heinze, Michael, 16, 169
Helsinki, 87
Heng, Janice, 97, 133, 161
Hernández Cordero, Adrián, 116, 161
Hill, Dan, 24, 161
Hines, Andy, 5, 28, 154, 161
Hobsbawm, Eric, 32, 161
Holiday Host, 60
Hollands, Robert G., 138, 161
HomeAway, 3, 24, 26, 75, 103, 107, 118, 124, 156, 164, 166
Hong Kong, 59, 135
Hong, Jose, 137, 161
Honigsberg, Peter Jan, 27, 28, 161
Horwath, 109, 161
Host Compliance, 97, 98, 124, 161
Hosteltur, 103, 161
hotel investment stop, moratorium, 88, 110, 112, 116–117, 120
Hotelschool The Hague, 1, 153, 160, 162, 169
HOTREC, 96, 161
housing market, 5, 25, 55, 88–90, 94, 98, 100, 105, 127, 132, 139, 145, 157

housing prices, 19, 91, 108, 111, 124, 159
Huet, Ellen, 61, 161
Huete, Raquel, 117, 161
Hurst, Nathan, 127, 161
Hyatt, 73, 168, 170

Iambnb, 50, 60
Iansiti, Marco, 29, 161
Ibiza, 86
Ikkala, Tapio, 3, 53, 161,
incremental demand, 4, 68
Indonesia, 134–136
Intel, 137, 138
International Labour Office, 61, 63, 154
Ioannides, Dimitri, 84, 85, 161
IPK, 68, 79, 161
Italy, 31, 32, 80, 81, 88, 130
iTunes, 22, 28

Jacobs, Thijs, 118, 162
Jacquot, Sébastien, 86, 160
Jakarta, 135
Japan, 38, 39, 103, 104, 156, 157
Jetblue, 34, 37
JLL, 135, 162
Jo, Yeo Sam, 97, 133, 161
Johnson, Bobbie, 129, 162
Joie de Vivre Hotels, 73
Joo, Daisy, 97, 162
Jullien, Bruno, 20, 155
Jun Sen, Ng, 137, 162
Juniper Research, 138, 162

Kagermeier, Andreas, 3, 162, 168
Kam Fung So, Kevin, 47, 162
Kanazawa, 38
Katharansiporn, Kanana, 97, 168
Katz, Lena, 71, 162
Kazaa, 28
KeyOkay, 60
Khoo, Lynette, 137, 162
Kiss, Jemima, 18, 123, 155
Kissimmee, 60
Kleenex, 4
Kleinherenbrink, Amber, 50, 162
Kloosterboer, Dirk, 97, 162
Kolar, Tomaz, 33, 162
Koomans, Marije, 64, 162
Kostakis, Vasilis, 11, 162
Kraniotis, Leen, 16, 162
KRO-NCRV, 107, 162

Kuala Lumpur, 135
Kusisto, Laura, 89, 162

Lakhani, Karim R., 29, 161
Lalicic, Lidiya, 48, 162
Lampinen, Airi, 3, 53, 161
Langford, Guy, 45, 163
Las Vegas, 32
Lee, Dayne, 89, 163
Lee, Terence, 139, 163
Lees, Loretta, 85, 163
Lehane, Chris, 102
Lehr, Dean D., 17, 163
'level playing field', 67, 70, 96, 106, 143
Lewis, P., 129, 163
Li, Jun, 58, 132, 163
Liang, Annabelle, 137, 163
Liang, Lena Jingen, 48, 163
Liang, Sai, 59, 173
Licata, Maria Cristna, 23, 155
Lim, Andriew, 28, 163
Lines, Greggary E., 96, 97, 99, 163
Linux, 11
Lisbon, 94
ListMinut.be, 61
Littman, Julie, 128, 163
'living like a local', 4, 31, 36, 40, 41, 46, 47, 130, 142
Lodder, Ton, 72, 163
Lomas, Natasha, 107, 163
London, 13–16, 26, 43, 55–58, 62–65, 68–70, 72, 75, 78, 80–82, 84, 85, 87, 97, 101, 125, 137, 143, 153, 155, 156, 159, 164, 165, 167
Lonely Planet, 46, 122, 130
Los Angeles, 26, 63, 86, 89, 125, 163, 166
low-cost, 79, 88, 120, 121, 147, 149, 170
Loy, Thomas, 97, 163
Lu, Jiaying, 48, 163
Lubbe Bakker, Sabine, 105, 163
Luca, Michael, 65, 158
Luxury Homes, 74
Lyft, 22, 62
Lyon, Katie, 89, 153

M3 Online Panel, 45
Macau, 135
MacCannel, Dean, 32, 163
Machu Picchu, 148

Madrid, 13–16, 55, 57, 58, 65, 68, 69, 71, 80, 83–87, 91, 98, 153, 159, 167
Mahmud, Aqil Haziq, 98, 136, 137, 163
Málaga, 91, 111, 156
Malaysia, 134–136
Malibu, 123
Mallorca, 98, 103, 117
Malone, Amelia R., 62, 158
management companies, 14, 59, 60, 160
Manhattan, 40, 89
Mantecón, Alejandro, 117, 161
Mao, Zhengxing, 48, 59, 163, 170
Marina Bay Sands, 134
Marriott, 67, 70, 155
Martin, C.J., 8, 163
Martins, Marco, 104
matchmaking, 4, 20, 21, 23–25, 28–30, 60, 139, 142, 159
Matera, 81
May, Kevin, 118, 162
mayi.com, 25
McAfee, Andrew, 16, 164
McCann, Chris, 125, 127, 164
McFarland, Henry B., 22, 23, 164
McGill University, 90, 170
Meek, James, 62, 145, 164
Meltzer, Josh, 67
Memorandum of Understanding Municipality of Amsterdam/Airbnb, 107
Merchiers, Jeroen, 52
Metro-Goldwyn-Meyer, 95
Meyer, Hans, 71
Miami, 26
Michel, Boris, 85, 159
Microsoft, 21, 23
Middel, Mark, 98, 155
Milan, 80, 85
Milikowski, Floor, 100, 164
millennials, 4, 40, 42, 44, 53, 60–62, 75, 120, 151, 157
Mishkin, S., 17, 157
Molz, Jennie Germann, 11, 164
Monaco, 55
Mondragón, 11
Moonies (Unification Church), 33, 34
Moragas, Miguel de, 114, 164
Morgan Stanley, 44, 49, 68, 70, 164
Mormons, 33
Morozov, Evgeny, 61, 148, 164
multi-homing, 21, 22, 24, 25

multilisters, 14–17, 54–55, 57, 59, 60, 62–64, 100, 115, 137, 139, 143, 146, 148, 149
Mundo, El, 117, 158
Mykonos, 90

Naafs, Saskia, 100, 164
Nam, Taewoo, 138, 164
Napster, 27, 28, 160, 161
NBTC Holland Marketing, 2, 164
Neirotti, Paolo, 138, 164
neoliberalism, 8, 93, 145–148, 160, 163
Netherlands, 54, 60, 160
New York, 17, 26, 39, 40, 43, 65, 68, 69, 72, 89, 90, 97, 98, 102, 103, 123–125, 137, 151, 153–155, 157–160, 162–164, 166–171
New York State Attorney General, 14, 55, 58, 89, 103, 164
New York Times, 46, 73, 154, 157, 158, 167, 168
New Zealand, 32, 134
Newgard Development Group, 60
Nicolau, Juan L., 59, 64, 170
Nieuwland, Shirley, 97, 164
Niido, 60
Norton, Steven, 16, 164
Novy, Johannes, 86, 160, 164
Nowak, Brian, 44, 49, 68, 70, 164
Nulty, Chrustopher, 18
Numbeo, 125

Oakland, 65, 123
Oasis, 73, 74, 168
Oates, Greg, 49, 165
Obama, Barack, 34
O'Connor, Peter, 23, 165
OECD, 20, 158, 166
'off the beaten track', 39, 43, 37, 49, 51, 80, 85, 111, 112, 142
Office of Short-Term Rentals, 99, 123
Olson, Kristine, 44, 165
Önder, Irem, 25, 47, 72, 160
onefinestay, 24–26, 72, 74, 170
O'Neill, John, 55, 165
online travel agents (OTAs), 16, 23, 24, 27, 30, 74–78, 152, 164
Oskam, Jeroen, 5, 11, 12, 19, 23, 24, 27, 58, 64, 76, 91, 111, 120, 132, 153, 157, 165

Ouyang, Yuxia, 55, 165
overtourism, 19, 79, 85, 88, 92, 110, 120, 143

País, El, 117, 121, 151, 153, 156, 158, 168, 169
Palo Alto, 123
Pan, Bing, 121, 165
Pardo, Theresa A., 138, 164
Paris, 12–18, 26, 36, 37, 40, 44, 50, 57, 58, 65, 68, 69, 72, 75, 80, 83–85, 87, 91, 94, 97, 104, 119, 125, 150, 151, 153, 155, 156, 158–160, 165
Parool, Het, 110, 165
Passiak, David, 31, 165
Paugh, Laura, 70
Peerby, 11
Peers, 35
Peltier, Dan, 54, 165
People's Action Party (PAP), 134
Perloth, Nicole, 129, 165
Pesonen, Juho, 45, 169
Peterson, Richard A., 32, 165
Pezenka, Ilona, 47, 165
Phoenix, 97
Picascia, Stefano, 80, 85, 165
Piketty, Thomas, 62
Pine, B. Joseph II, 33, 39, 159
Pinkster, Fenne M., 112, 165
Playa de las Catedrales, 88
Poon, Linda, 139, 165
Pop-Tarts, 4
Port of Amsterdam, 109, 165
Porter, Michael E., 23, 165
Portland, 17, 52, 53, 151, 159
Portugal, 52, 98, 146
Prague, 87
Priceline, 23
privacy protection, 17, 93, 94, 98, 99, 102, 139, 144
privatisation, 116, 145, 148
professionalisation, 16, 58–60, 64, 132, 143
Proposition F, 17, 123, 155
PUBLIC, 72
Puebla, 39, 40
PWC, 69, 109, 160

Quattrone, Giovanni, 64, 65, 81, 98, 99, 166
Queens, 40

racial discrimination, 65, 66, 96, 104, 143, 158
Ranger, Terence, 32, 161
rate parity, 23, 25
Realty Shares, 60, 166
Redeker Sellner Dahs, 101, 166
registration systems, 96, 99, 101, 103, 104, 117, 118, 123, 124, 130, 131, 137, 152, 166
regulatory capture, 96
reintermediation, 23, 75
Rentalia, 103
ResonanceCo, 44, 166
Resort World Sentosa, 134
reviews, 24, 47, 48, 54, 59, 73, 74, 90, 95, 100, 159
Reykjavík, 18, 88, 91, 99, 153
Richard, Brendan, 72, 166
Rifkin, Jeremy, 9, 147, 166
Rijk, Mirjam de, 145, 166
Rochet, Jean-Charles, 20, 21, 23, 24, 166
Rodríguez-Planas, Nuria, 116, 153
Rogers, Roo, 10, 43, 155
Rome, 26, 55, 81
Romero, Víctor, 103, 166
Room Mate, 71, 73, 154, 158, 162
Rosen, Kenneth T., 89, 166
Rosenthal, Helen, 102
Roth, Howard, 45, 166
Rotterdam, 2, 64, 153, 162

safety regulations, 17, 67, 70, 94–97, 99, 104
Said, Carolyn, 123, 124, 131, 166
Sala i Martin, Xavier, 117, 166
Samaan, Roy, 63, 89, 166
Sampson, Hannah, 49, 166
San Francisco, 17, 18, 26, 29, 34, 65, 80, 89, 97, 99, 103, 122–125, 127–132, 144, 149, 152, 155, 156, 158–164, 167–169
San Francisco Residential Rent Stabilization and Arbitration Board, 127, 128
San Jose, 123
Santa Clara, 123
Santa Monica, 97, 157
Santorini, 86, 88, 90, 148, 167
Sanya, 135
Sarasola, Kike, 71, 73, 158, 162
Satama, Sampo, 48, 167

Saturn, 33
Schaal, Dennis, 75, 77, 167
Scheiber, Noam, 58, 169
Schmalensee, Richard, 20, 157
Schnitzler, Hans, 145, 167
Schrager, Ian, 72
Schumpeter, Joseph A., 46, 167
scraped data, 1, 2, 16, 55, 87, 99–101, 108
Second Love, 146, 150
self-actualisation, 32, 34, 36, 42, 53
self-regulation, 16, 17, 93, 100, 118
Servas, 129, 169
Servihabitat, 91, 167
Shanghai, 135
ShareNL, 105, 107, 152, 167
Sharetribe, 45
Sharing City, 1, 105, 107, 167
Shaw, Lucas, 28, 167
Shaw, Randy, 41, 167
Sheivachman, Andrew, 49, 167
Siguaw, Julia, 23, 156
Silicon Valley, 35, 107, 127, 128, 155
Singapore, 97, 98, 125, 133–140, 144, 152, 153, 156, 160–163, 165, 167–170
Singapore Legal Advice, 137, 167
Singapore Smart Nation, 134, 167
Singapore Tourism Board, 135, 136, 140, 167
Skift, 52, 75, 77, 156, 165–170
Slee, Tom, 1, 12, 14, 17, 35, 54, 55, 58, 64, 93, 100, 103, 157, 167
Slevigen, Henrik Hammerhei, 48, 170
Smith, Alison Jane, 148, 167
Smith, Helena, 91, 167
Smith, Neil, 85, 167
Smith, Noah, 145, 167
Snappcar, 11
Sondermeijer, Vincent, 2, 167
Sonoma, 73, 168
Sony, 27
Sorenson, Arne, 67
Soriano, Domingo, 117, 167
Sottek, T.C., 17, 18, 167
Spain, 44, 52, 80, 91, 114, 130, 145, 146, 155, 168
spatial concentration, 64, 79–83, 158, 160
Spotify, 22, 28, 167
Spotswood, Beth, 128, 168
Sritama, Suchat, 97, 168
standardisation, 28, 73, 143, 144, 149

Staten Island, 40
Statista, 69
Stern, Joseph, 24, 168
Stockholm, 87
Stors, Natalie, 3, 162, 168
STR, 69, 70, 160
Strong, Colin, 67, 168
Sundararajan, Arun, 8, 17, 61, 62, 95, 96, 168
Sunderland, Judith, 145, 168
superhost, 55, 58–60, 73, 74, 152, 160, 163
Sussman, Sam, 134, 168
Sweden, 134

Taiwan, 134
TakeLessons, 61
Tallinn, 55, 66
Tam, Donna, 123, 168
Taskrabbit, 61
taxes, taxation, 2, 4, 9, 12, 16–20, 52, 67, 70, 88, 94, 96–99, 104, 106, 107, 119, 123, 131, 143, 144, 152, 157, 159, 160
Taylor, John, 32, 168
TCP/IP, 29, 30
Thailand, 97
The Friendly Host, 60
The Venetian, 31
Thuisafgehaald, 11
Tim, Ho Chi, 134, 168
Ting, Deanna, 70, 73–75, 101, 167–169
Tirole, Jean, 20, 21, 23, 24, 166
Tokyo, 38, 161, 170
tourist tax, 2, 17, 88, 94, 99, 106, 107, 123, 157
touristification, 19, 32, 84, 85, 90, 91, 110–112, 115, 116, 124, 145, 147, 148, 154
Trading Economics, 134, 169
transparency, 6, 16, 17, 19, 29, 43, 88, 100, 102–104, 107, 119, 123, 131, 139, 144, 160
Trulia, 124, 169
Trump, Donald, 121, 128
Tucker, Catherine, 22, 169
Tujia, 25, 26, 74
Turnbull, David, 16, 169
Turo, 62
Tussyadiah, Iis, 44, 45, 169
two-sided markets, 20, 23, 26, 75, 152, 158, 164–166

Uber, 3, 10, 11, 17, 18, 20, 22, 58, 61, 63, 69, 94–96, 145, 158, 161, 167, 170
UK Office for National Statistics, 16
Unification Church ('Moonies'), 33, 34
United States of America, 45, 90, 94, 98, 117, 128, 129
Universal, 27
UNWTO, 79, 108, 109, 169
Urban Redevelopment Authority (URA), 136–138, 163, 169, 170
Utrecht, 84, 85, 161

Valencia, 98, 103, 155, 166
Van Burgeler, Marieke, 36, 169
Van den Bent, Cato, 50, 169
Van den Broek, Vera, 51, 169
Van der Bijl, Vincent, 90, 169
Van der Heijden, Kees, 5, 169
Van der Most, Kees, 2, 170
Van Loenen, Marlot, 49, 169
Vanguardia, La, 103, 118, 163
Vekshin, Alison, 17, 169
Venema, Aart-Jan, 145, 166
Venice, 31, 32, 81, 86, 87, 110, 111, 145, 148, 155,
Verdú, Daniel, 105, 169
Verwaltungsgericht Berlin, 101, 169
Vienna, 25, 47, 75, 76, 160
Visjø, Camilla Tryggestad, 48, 170
Visser, Barbara, 51, 170
Volkswagen, 33
Vranken, Jos, 2, 170

Wachsmuth, David, 90, 170
Wallaart & Kusse Public Affairs, 18
Wallsten, Scott, 69, 170
Wang, Dan, 59, 64, 170
Wang, Ning, 32, 33, 170
Ward, Jill, 16, 170
Warner, 27
Warriner, John, 45, 156
Washington, 125
Wei, Low De, 137, 170
Weismayer, Christian, 48, 162
Welch, Liz, 71, 72, 170
Wen-Yi, Lee, 137, 170
WeWork, 71, 159
Whyte, Patrick, 73, 146, 170
Wikipedia, 11
Wilczynski, Jake, 102

Wimdu, 3, 26, 107
Windows, 22
winner-takes-all competition, 4, 21, 22, 28, 75, 119, 130, 142
World Bank Group, 134, 170
Wright, Julian, 21, 152

Xiang, Yixiao, 26, 170
Xiang, Zheng, 16, 170
Xiaozhu, 25, 26, 156
Xie, Karen, 59, 170

Yang, Sujin, 47, 170
Yannopoulou, Natalia, 24, 170
Yeoman, Ian S., 33, 160, 170

Youtube, 36–39, 41, 151, 152
Yuan, Lin Yee, 72, 170

Zabkar, Vesna, 33, 162
Zagreb, 55
Zaleski, Olivia, 60, 74, 171
Zandberg, Tjeerd, 5, 23, 24, 27, 76, 165
Zervas, Georgios, 2, 48, 69, 171
Zillow, 125, 126
Zipcar, 9
Zoku, 71, 72, 159
Zubizarreta, Iker, 138, 171
Zukin, Sharon, 85, 171
Zweckentfremdung (purpose alienation), 100, 166, 169